Praise for James Hansen and *Storms of My Grandchildren*

"Rich in invaluable insights into the geopolitics as well as the geophysics of climate change, Hansen's guaranteed-to-be-controversial manifesto is the most comprehensible, realistic, and courageous call to prevent climate change yet."

—*Booklist* (starred)

"After sounding the climate alarm in papers and conferences for two decades, here Hansen takes off the gloves . . . With urgency and authority, Hansen urges readers to speak out—taking to the streets if necessary—to protect the Earth from calamity for the sakes of their children and grandchildren."

—*Kirkus* (starred)

"An urgent and provocative call to action and the full story of what we need to know about humanity's last chance to get off the path to a catastrophic meltdown, and why we don't know the half of it. While the truth may be uncomfortable, it is also exciting to learn so much that is essential . . . It is number one on my list for both myself and a few friends."

—*Sierra Atlantic*

"Dr. James Hansen is Paul Revere to the foreboding tyranny of climate chaos—a modern-day hero who has braved criticism and censure and put his career and fortune at stake to issue the call to arms against the apocalyptic forces of ignorance and greed."

—**Robert F. Kennedy Jr.**

STORMS
OF MY
GRANDCHILDREN

The Truth About the Coming Climate Catastrophe
and Our Last Chance to Save Humanity

JAMES HANSEN

Illustrations by Makiko Sato

BLOOMSBURY
New York · Berlin · London · Sydney

Published by Bloomsbury USA, New York

All papers used by Bloomsbury USA are natural, recyclable products made from
wood grown in well-managed forests. The manufacturing processes conform to
the environmental regulations of the country of origin.

LIBRARY OF CONGRESS CATALOGING-IN-PUBLICATION DATA

Hansen, James C.
Storms of my grandchildren : the truth about the coming climate catastrophe
and our last chance to save humanity / James Hansen ; illustrations by
Makiko Sato.—1st U.S. ed.
p. cm.
Includes bibliographical references and index.
ISBN 978-1-60819-200-7 (alk. paper hardcover)
1. Global warming. 2. Climatic changes. 3. Greenhouse gases. 4. Human
beings—Effect of climate on. I. Title.
QC981.8.G56H365 2009
363.738'74—dc22
2009044553

First published by Bloomsbury USA in 2009
This paperback edition published in 2011

Paperback ISBN: 978-1-60819-502-2

1 3 5 7 9 10 8 6 4 2

Designed by Rachel Reiss
Typeset by Westchester Book Group
Printed in the U.S.A by Quad/Graphics, Fairfield, Pennsylvania

To Sophie, Connor, Jake,
and all the world's grandchildren

Contents

Preface

Planet Earth, creation, the world in which civilization developed, the world with climate patterns that we know and stable shorelines, is in imminent peril. The urgency of the situation crystallized only in the past few years. We now have clear evidence of the crisis, provided by increasingly detailed information about how Earth responded to perturbing forces during its history (very sensitively, with some lag caused by the inertia of massive oceans) and by observations of changes that are beginning to occur around the globe in response to ongoing climate change. The startling conclusion is that continued exploitation of all fossil fuels on Earth threatens not only the other millions of species on the planet but also the survival of humanity itself—and the timetable is shorter than we thought.

How can we be on the precipice of such consequences while local climate change remains small compared with day-to-day weather fluctuations? The urgency derives from the nearness of climate tipping points, beyond which climate dynamics can cause rapid changes out of humanity's control. Tipping points occur because of amplifying feedbacks—as when a microphone is placed too close to a speaker, which amplifies any little sound picked up by the microphone, which then picks up the amplification, which is again picked up by the speaker, until very quickly the noise becomes unbearable. Climate-related feedbacks include loss of Arctic sea ice, melting ice sheets and glaciers, and release of frozen methane as tundra melts. These and other science matters will be clarified in due course.

There is a social matter that contributes equally to the crisis: government greenwash. I was startled, while plotting data, to see the vast disparity between government words and reality. Greenwashing, expressing concern about global warming and the environment while taking no actions to actually stabilize climate or preserve the environment, is prevalent in the United States and other countries, even those presumed to be the "greenest."

The tragedy is that the actions needed to stabilize climate, which I will describe, are not only feasible but provide additional benefits as well. How can it be that necessary actions are not taken? It is easy to suggest explanations—the power of special interests on our governments, the short election cycles that diminish concern about long-term consequences—but I will leave that for the reader to assess, based on the facts that I will present.

My role is that of a witness, not a preacher. Writer Robert Pool came to that conclusion when he used those religious metaphors in an article about Steve Schneider (a preacher) and me in the May 11, 1990, issue of *Science*. Pool defined a witness as "someone who believes he has information so important that he cannot keep silent."

I am aware of claims that I have become a preacher in recent years. That is not correct. Something did change, though. I realized that I am a witness not only to what is happening in our climate system, but also to greenwash. Politicians are happy if scientists provide information and then go away and shut up. But science and policy cannot be divorced. What I've seen is that politicians often adopt policies that are merely convenient—but that, using readily available scientific data and empirical information, can be shown to be inconsistent with long-term success.

I believe the biggest obstacle to solving global warming is the role of money in politics, the undue sway of special interests. "But the influence of special interests is impossible to stop," you say. It had better not be. But the public, and young people in particular, will need to get involved in a major way.

"What?" you say. You already did get involved by working your tail off to help elect President Barack Obama. Sure, I (a registered Independent who has voted for both Republicans and Democrats over the years) voted for change too, and I had moist eyes during his Election Day speech in Chicago. That was and always will be a great day for America. But let me tell you: President Obama does not get it. He and his key advisers are subject to heavy pressures, and so far the approach has been, "Let's compromise." So you still have a hell of a lot of work ahead of you. You do not have any choice. Your attitude must be "Yes, we can."

I am sorry to say that most of what politicians are doing on the climate front is greenwashing—their proposals sound good, but they are deceiving you and themselves at the same time. Politicians think that if matters look difficult, compromise is a good approach. Unfortunately, nature and the laws of physics cannot compromise—they are what they are.

Policy decisions on climate change are being deliberated every day by those without full knowledge of the science, and often with intentional misinformation spawned by special interests. This book was written to help rectify this situation. Citizens with a special interest—in their loved ones—need to become familiar with the science, exercise their democratic rights, and pay attention to politicians' decisions. Otherwise, it seems, short-term special interests will hold sway in capitals around the world—and we are running out of time.

My approach in this book is to describe my experiences as a scientist interacting with policy makers over the past eight years, beginning on my sixtieth birthday in 2001, the day I spoke to Vice President Dick Cheney and the cabinet-level Climate Task Force. Each chapter discusses a facet of climate science that I hope a nonscientist will find easy to understand. Chapter 1 may be the most challenging. It discusses climate forcing agents—or simply, climate forcings—the subject of my presentation to the Task Force.

The official definition of a climate forcing may seem formidable: "an imposed perturbation of the planet's energy balance that tends to alter global temperature." Examples make it easier: If the sun becomes brighter, it is a climate forcing that would tend to make Earth warmer. A human-made change of atmospheric composition is also a climate forcing.

In 2001 I was more sanguine about the climate situation. It seemed that the climate impacts might be tolerable if the atmospheric carbon dioxide amount was kept at a level not exceeding 450 parts per million (ppm; thus 450 ppm is 0.045 percent of the molecules in the air). So far, humans have caused carbon dioxide to increase from 280 ppm in 1750 to 387 ppm in 2009.

During the past few years, however, it has become clear that 387 ppm is already in the dangerous range. It's crucial that we immediately recognize the need to reduce atmospheric carbon dioxide to at most 350 ppm in order to avoid disasters for coming generations. Such a reduction is still practical, but just barely. It requires a

*My first grandchild, at almost two years
old—changing my perception.*

prompt phaseout of coal emissions, plus improved forestry and agricultural practices. That part of the story will unfold in later chapters, but we need to acknowledge now that a change of direction is urgent. This is our last chance.

I myself changed over the past eight years, especially after my wife, Anniek, and I had our first grandchildren. At the beginning of this period, I sometimes showed a viewgraph of the photo of our first grandchild on this page during my talks on global warming. At first it was partly a joke, as newspapers were referring to me as the "grandfather of global warming," and partly pride in a young lady who had become an angel in our lives. But gradually, my perception of being a "witness" changed, leading to a hard decision: I did not want my grandchildren, someday in the future, to look back and say, "Opa understood what was happening, but he did not make it clear."

That resolve was needed. If it hadn't been for my grandchildren and my knowledge of what they would face, I would have stayed focused on the pure science, and not persisted in pointing out its relevance to policy. When policy is brought into the discussion, it seems that a lot of forces begin to react. I prefer to just do science. It's more pleasant, especially when you are having some success in your investigations. If I must serve as a witness, I intend to testify and then get back to the laboratory, where I am comfortable. That is what I intend to do when this book is finished.

BECAUSE THE BOOK opens on my sixtieth birthday, I should mention here a bit about where I came from. I was lucky to be born in a time and place—in Iowa, where I was in high school when Sputnik was launched—that I could be introduced to science in a way that seemed to be normal, yet was very special.

I grew up in western Iowa, one of seven children. My father was a tenant farmer, educated only through the eighth grade, and my parents divorced when I was young. But in those days a public college was not expensive, so it was pretty easy for me to save enough money to go to the University of Iowa.

My career in science, my first step into science research, was born one evening in December 1963. The day before, fellow student Andy Lacis and I had swept leaves, cobwebs, and mice out of a little domed building on a hill in a cornfield just outside Iowa City. The next night, within that dome, an older graduate student, John Zink, helped us use a small telescope to observe a lunar eclipse. When the moon went into eclipse, passing into Earth's shadow, we were surprised that we saw nothing—just a black area in the sky, without stars, in the spot where we had just seen a full moon; the moon had become invisible to the naked eye. This is not usually the case with an eclipse. Normally, the moon is dimmed but still obvious, because sunlight is refracted by Earth's atmosphere into the shadow region. However, nine months earlier, in March 1963, there had been a large volcanic eruption, of Mount Agung on the island of Bali, which injected sulfur dioxide gas and dust into Earth's stratosphere. The sulfur dioxide gas combined with oxygen and water to form a sulfuric acid haze, and the resulting particles in the stratosphere blocked most of the sunlight that normally is refracted into Earth's shadow.

We measured the brightness of the moon with a photometer attached to the telescope, and in the next year I was able to figure out how much material there must have been in Earth's stratosphere to make the moon as dark as it was. Mainly that required reading some papers (in German) written by the Czechoslovakian astronomer

František Link, who had worked out the equations for the eclipse geometry, and writing a computer program for the calculations. The result was my first scientific paper, published in the *Journal of Geophysical Research*—my first experience as a witness, at least a witness of science, if not in the biblical sense.

Our good fortune was that we had found our way into the Physics and Astronomy Department of a remarkable man, James Van Allen. An astronomy professor in Van Allen's department, noticing that Andy and I were capable students, convinced us to take the physics graduate school qualifying examinations in our senior year. We were the first undergraduates to pass that exam, and, perhaps as a result, we both were offered NASA graduate traineeships, which fully covered our costs to attend graduate school.

I was so shy and uncertain of my abilities that I had avoided taking any of Professor Van Allen's classes, not wanting to reveal my ignorance. But Van Allen noticed me anyhow—probably because I had not only passed the graduate exam but also received one of the higher scores. He told me about recent observational data concerning the planet Venus, which suggested that either the surface of Venus must be very hot or the planet had a highly charged ionosphere emitting microwave radiation. When I started to work on the Venus data for a Ph.D. thesis, Van Allen appointed himself as chairman of my thesis committee. If it had not been for the attentiveness and generosity of this soft-spoken, gentle man, whom no student ever should have been intimidated by, I probably would not have gotten involved in planetary studies.

More than a decade later, in 1978, I was still studying Venus. And by then I was responsible for an experiment that was on its way to that planet, aboard the Pioneer Venus mission. In the five years since I had proposed that experiment to measure the properties of the Venus clouds, I had been working about eighty hours per week. Anniek, whom I had met while I was on a postdoctoral fellowship at the University of Leiden Observatory in the Netherlands, continued to believe me, each year, when I said that the next year I would have more time. Then I had to tell her that, after all that effort, I was going to resign from the Pioneer mission before it arrived at Venus, turning the experiment over to Larry Travis, another friend and colleague from Iowa.

The reason: The composition of the atmosphere of our home

planet was changing before our eyes, and it was changing more and more rapidly. Surely that would affect Earth's climate. The most important change was the level of carbon dioxide, which was being added to the air by the burning of fossil fuels. We knew that carbon dioxide determined the climate on Mars and Venus. I decided it would be more useful and interesting to try to help understand how the climate on our own planet would change, rather than study the veil of clouds shrouding Venus. Building a computer model for Earth's climate was also going to be a lot more work. As always, Anniek accepted, and tried to believe, my promise that it would be a temporary obsession.

Another decade later, on June 23, 1988, I was a witness, an official witness, when I testified to a Senate committee chaired by Tim Wirth of Colorado. I declared, with 99 percent confidence, that it was time to stop waffling: Earth was being affected by human-made greenhouse gases, and the planet had entered a period of long-term warming. Combined with an unusually hot and dry summer and the attention global warming was getting nationally and internationally, my announcement garnered broad notice.

It soon became apparent, though, that my testimony, combined with the weather, was creating a misimpression. Global warming does increase the intensity of droughts and heat waves, and thus the area of forest fires. However, because a warmer atmosphere holds more water vapor, global warming must also increase the intensity of the other extreme of the hydrologic cycle — meaning heavier rains, more extreme floods, and more intense storms driven by latent heat, including thunderstorms, tornadoes, and tropical storms. I realized that I should have emphasized more strongly that both extremes increase with global warming.

Therefore I sought one more opportunity to be a witness. Senator Al Gore provided that opportunity at a hearing in the spring of 1989. When I sent Senator Gore a note before the hearing, explaining that my written testimony had been altered by the White House Office of Management and Budget to make my conclusions about the dangers of global warming appear uncertain, he alerted the media, assuring that there would be widespread coverage of the testimony. Unfortunately, the message about the wet end of the hydrologic cycle was lost in the brouhaha. Mother Nature, however, responded four years later with a "hundred-year" flood, one that normally occurs only once a century, which submerged Iowa and much of the Midwest. They were hit with another "hundred-year" flood in 2008.

After my testimony at Gore's hearing, I was firmly resolved to go back to pure science and leave media interactions to people such as Steve Schneider and Michael Oppenheimer, people who were more articulate and seemed to enjoy the process. But after another decade I made an exception and agreed to debates in 1998 with the global warming "contrarians" Dick Lindzen and Pat Michaels, because I had a clear scientific purpose: I wanted to present and publish a table of the key differences between my position regarding global warming and the position of the contrarians. My expectation was that the table's specificity would permit future evaluation of our positions. I would use this table in my meeting with Vice President Cheney's Climate Task Force in 2001.

So for more than a decade after the Gore hearing in 1989, I was able to stick strictly to science, turning down many opportunities to appear on documentaries and other television programs. It was that science that I would discuss with the Climate Task Force.

The [obscured]
Cli[obscured]

Albritton, director of the [obscured]
ministration (NOAA) Aer[obscured]
and about my age, was [obscured]
been NOAA's chief s[obscured]
of human-made ga[obscured]
decade or so you[obscured]
cal Fluid Dyn[obscured]
models are [obscured]
surface, a[obscured]
ics of th[obscured]
Th[obscured]
tha[obscured]

A LARGE POLICE DOG WAS [obscured] sumed that it was checking for bomb[obscured] was about to arrive. This was the first meeting [obscured] Climate Task Force. It fell on my sixtieth birthday, March 2[obscured].

"Climate Working Group" was the phrase I remembered from the phoned invitation, so that was the title I put on the handout I brought to the meeting. There was no letter of invitation, and I was not given any paperwork at the meeting. Later, President George W. Bush and the media referred to this group as the Climate Task Force, so I will use that title here.

The Climate Task Force consisted of six cabinet members plus the national security adviser (Condoleezza Rice), the EPA administrator (Christine Todd Whitman), and Vice President Cheney as chairman. The cabinet members were Secretary of State Colin Powell, Spencer Abraham (Energy), Paul O'Neill (Treasury), Gale Norton (Interior), Ann Veneman (Agriculture), and Donald Evans (Commerce).

We were three scientists who had been requested to explain the current understanding of climate change and the role that humans might have in causing global warming. We were a bit nervous. This was surely the most high-powered group that any of us had spoken to.

When I arrived early that morning at the Department of Commerce headquarters in Washington, D.C., the venue for the meeting, I found the other two scientists, Dan Albritton and Ron Stouffer, hunched over a table looking at the charts Ron planned to show. Dan was advising Ron to reduce the amount of complicated material.

National Oceanic and Atmospheric Ad-
onomy Laboratory in Boulder, Colorado,
an old hand at presentations. He had long
okesman for describing research on the effect
es on the stratospheric ozone layer. Stouffer, a
ger, is a climate modeler at the NOAA Geophysi-
mics Laboratory in Princeton, New Jersey. Climate
omputer simulations of the atmosphere, ocean, land
d their interactions, which are used to study the dynam-
climate system and how future climate may change.

backdrop for this meeting was President Bush's confirmation
the United States would not sign the Kyoto Protocol. "Kyoto"
quired developed countries to reduce emissions of human-made
heat-absorbing greenhouse gases to several percent below 1990 emission rates. The main greenhouse gas is carbon dioxide, which is increasing because of the burning of fossil fuels: coal, oil, and gas.
Deforestation contributes a smaller amount, about 20 percent, to the
carbon dioxide increase.

The president's refusal to sign on to Kyoto was expected. More
important was the revelation on March 13 that the United States
would not regulate carbon dioxide emissions from power plants.
That decision was a heavy blow to environmentalists and scientists who realized that Earth's climate was approaching a dangerous
situation because of the buildup of atmospheric carbon dioxide.

Coal burning at power plants is the greatest source of increasing
atmospheric carbon dioxide. It is also the source most susceptible
to control. The decision not to restrict power plant emissions re-
neged on a promise Bush made repeatedly during the 2000 presidential election campaign.

Bush had pledged to include carbon dioxide in a "four pollutant
strategy" to reduce the most damaging pollutants from power plants.
That promise, together with the Clinton-Gore administration's poor
record in constraining carbon dioxide emissions, stymied Al Gore
from raising the environment and climate change as an effective
campaign issue. Given the razor-thin margin in the 2000 election,
and the environmental awareness of Florida voters, it seems clear
that Gore would have become president if it were not for Bush's
pollution-reduction promise.

Despite that backdrop, the fact that the Bush-Cheney administration was having these Task Force meetings suggested that it took the climate change issue seriously and wanted to learn more about it. The implication was that future policies were still open and could be influenced by scientific evidence.

The vice president abetted these impressions in his opening remarks and noted that Task Force meetings would be "principals only"—participants could not send representatives in their stead. Treasury Secretary O'Neill related that the previous afternoon he had met with President Bush, who had said that he wanted the United States to take a leadership role in addressing climate change, even though it would need to be via a route other than the Kyoto Protocol.

Colin Powell apologized that he would need to step out during this meeting for a phone discussion with Yasir Arafat. Rats, I thought—that was disappointing. I had tailored some of my planned remarks for the secretary of state.

I was hopeful that there was a good chance that the group as a whole would favor actions to stem climate change. Powell and Rice surely felt the anger of Europe, Japan, and other nations about the failure of the United States to ratify the Kyoto Protocol. They must also have realized the benefit, for national security reasons, of reducing our dependence on fossil fuels. Both the Treasury's O'Neill and the EPA's Whitman had made speeches about the dangers of global warming and the need for strong policies to reduce fossil fuel emissions.

On the other side, Energy Secretary Abraham had stated in a public speech on March 19 that the United States must add ninety new power plants each year, mostly coal-fired, for the next twenty years to meet the need for a 45 percent increase in electricity demand by 2020. Vice President Cheney strongly supported efforts to increase fossil fuel supplies, including the opening of public lands, continental shelves, and the Arctic for increased coal mining and oil and gas drilling.

Altogether it was unclear where the balance of opinion of the Task Force would fall. I thought it was realistic to think the scientific information we provided would aid their decision making.

Dan Albritton gave an overview called "Climate Change: What We Know and What We Don't." Each of his hand-drawn charts included a little "thermometer," or confidence index, with fluid rising to a level between zero and ten. Our understanding that the natural greenhouse effect keeps Earth much warmer than it would be otherwise rated a "ten," for example, while our ability to describe regional effects of global warming rated only a "three."

The basis for Albritton's presentation was a set of reports by the United Nations Intergovernmental Panel on Climate Change (IPCC), which was in the process of being published at that time. Albritton's bottom line, consistent with the rigorous position of IPCC, was that the scientific community provided policy-relevant information but would make no statements about policy.

My presentation was titled "The Forcing Agents Underlying Climate Change." Forcing agents are factors that affect the energy balance and temperature of Earth, such as carbon dioxide in the atmosphere. I began by noting that climate, the average weather over a finite interval, fluctuates without any forcing, because the atmosphere and ocean are chaotic fluids that are always sloshing about. There is no way to predict far ahead of time where and how big any particular slosh will be. But once we see a slosh coming, we can project how it will play out. That is what weather forecasting consists of—mapping the current sloshes and looking upstream at where the next ones are coming from. In winter in the United States, if the next slosh is coming straight down from Canada, watch out!

Yet, despite this unpredictable sloshing, if weather is averaged over a long enough time, the system is "deterministic," that is, it responds to a forcing mechanism in a predictable way. For example, if we begin to slowly move Earth closer to the sun, we can be sure that Earth will become warmer. Not every year though, because chaos and sloshings also occur on greater time and space scales than local day-to-day weather.

One big, slow sloshing on Earth is the "Southern Oscillation," in which surface waters of the Pacific Ocean at the equator oscillate between warm El Niño and cool La Niña phases. La Niña is caused by the upwelling of a large amount of water from the cold, deep ocean along the South American coast. The effect of El Niño or La Niña on global temperature and precipitation patterns is huge. This oscillation would make it hard to notice immediately an underlying global warming trend due to Earth moving closer to the sun, for instance. The ocean would eventually get warmer, but that would take time because the ocean is two and a half miles deep.

Of course, the subject of interest to the Task Force was not the hypothetical case of Earth being moved closer to the sun. Rather, I needed to discuss real climate forcing mechanisms, some of which were well measured and others only crudely estimated.

I defined a climate forcing as an imposed perturbation (disturbance) of the planet's energy balance. It is measured in watts per square meter. For example, if the sun becomes 1 percent brighter, that is a forcing of about two watts (for brevity I sometimes will omit "per square meter" in discussing forcings), because Earth absorbs about 240 watts of sunlight averaged over day and night.

One large climate forcing that we know about is caused by volcanic eruptions that inject sulfur dioxide gas into Earth's lower stratosphere (altitude ten to twenty miles). Sulfur dioxide combines with oxygen and water to form tiny sulfuric acid droplets (aerosols) that scatter sunlight back to space, reducing solar heating of Earth's surface. Aerosols created by the 1991 eruption of Mount Pinatubo in the Philippines reduced solar heating of Earth by almost 2 percent, a negative forcing of about −4 watts. This large forcing, however, was present only briefly—after two years most of the Pinatubo aerosols had fallen out of the atmosphere. This brevity greatly reduces the effect of volcanoes on long-term climate, but an effect on the climate trend might be detectable if, say, there is an unusual concentration of volcanic eruptions in a given century.

The largest human-made climate forcing is due to greenhouse gases. These are gases that partially absorb infrared (heat) radiation, so an increased gas amount makes the atmosphere more opaque at infrared wavelengths. This increased opacity causes heat radiated to space to arise from a higher level in the atmosphere, where it is colder. Heat radiation to space is therefore reduced, resulting in a planetary energy imbalance. So Earth radiates less energy than it absorbs, causing the planet to warm up.

How much climate responds to a specified forcing—specifically, how much global temperature will change—is called "climate sensitivity." I told the Task Force that climate sensitivity is reasonably well understood, on the basis of Earth's history. Paleoclimate (ancient climate) records show accurately how Earth responded to climate forcings over the past several hundred thousand years.

However, because our presentations were limited to about twenty minutes each, I chose to focus on a comparison of the different climate forcing agents that drive climate change, as an effective way to show the human contribution to global warming. It is useful to simply compare the forcings, because the global temperature change is expected to depend on the size of the forcing, more or less independent of the forcing mechanism. This expectation is supported by climate model studies and empirical data (see chapter 3).

I showed the Climate Task Force a bar graph (figure 1) estimating all known climate forcings in 2000 relative to the beginning of the industrial revolution. The vertical lines (whiskers) represent the estimated uncertainties for each forcing.

The first seven forcings in figure 1 are all human-induced. However, it's useful to first discuss the natural forcings, changes of the sun and volcanic activity.

Changes of the sun's irradiance (brightness seen from Earth) cause a potentially significant climate forcing on decade-to-century time scales. Unfortunately, precise measurements of solar brightness became possible only with satellite observations that began in the late 1970s. These data revealed a cyclic change of about 0.1 percent with the ten-to-twelve-year solar magnetic cycle, yielding a ten-to-twelve-year cyclic forcing of just over 0.2 watts.

The direct effect of solar brightness is amplified by at least one indirect effect. The solar variability is much larger at ultraviolet

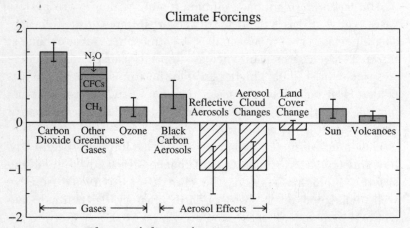

FIGURE I. *Change of climate forcings, in watts per square meter, between 1750 and 2000. Vertical bars show estimated uncertainty. Uncertainty for "other greenhouse gases" is similar to that for carbon dioxide. (Data from Hansen et al., "Efficacy of Climate Forcings." See sources.)*

wavelengths than it is at visible wavelengths. Ultraviolet radiation breaks up oxygen molecules in Earth's atmosphere, creating ozone, which increases the greenhouse effect. This indirect climate forcing enhances the direct solar forcing by perhaps as much as one third, making the total cyclic solar forcing about 0.3 watts.

Some who say the sun plays a larger role in climate change than carbon dioxide or other greenhouse gases hypothesize that there must be other indirect effects that magnify the small measured variations of solar brightness. The most common hypothesis is an almost Rube Goldberg concoction: the sun altering cosmic rays, which then alter cloud condensation nuclei, which alter cloud cover, which alters absorbed sunlight, which alters climate. However, there is no meaningful evidence supporting a large indirect amplification. The small cyclic component of global temperature that is extracted from statistical analyses of observed global temperature is consistent with a solar forcing of 0.2 to 0.3 watts. Possible errors in extracting the cyclic temperature response allow, at most, amplification of the solar forcing by a factor of two, which still leaves the cyclic solar forcing much smaller than the greenhouse gas forcing.

A bigger question about the sun remains: How large are solar variations on the century timescale? The 1750–2000 solar forcing estimated in figure 1 is based on research of solar experts, especially Judith Lean and Claus Fröhlich, who used indirect ("proxy") indicators of solar activity, such as sunspots. They assume that the relation between solar activity and solar brightness observed in the past few decades is the same as it was a few hundred years ago. That leads to the conclusion that recent solar forcing is a few tenths of a watt greater than it was in the eighteenth century. The uncertainty, however, is large, as figure 1 indicates.

The other known natural climate forcing mechanism, volcanoes, probably worked in the same sense as the sun over the interval from the mid-eighteenth to the mid-twentieth centuries. That is because the available data, meager as they are, suggest that volcanic activity was greater in the eighteenth century than in the twentieth century. Actually, between 1963 and 1991 three large volcanoes (Mount Agung, El Chichón, and Mount Pinatubo) erupted, a degree of volcanic activity that would have been at least comparable to that in the eighteenth century. However, when people compare eighteenth- and twentieth-century climates to examine the effects of natural climate forcings, they usually exclude the last few decades of the twentieth century because, by that time, the human-made greenhouse gas forcing was so large that it eclipsed the effects of even three large volcanic eruptions; whereas up through the middle of the twentieth century, the net human-made climate forcing was rather small.

The increase of climate forcing in the mid-twentieth century due to the changing level of volcanic activity, relative to the eighteenth century, is estimated as 0.15 watts, with an uncertainty of about 0.1 watts. Thus the net change of natural climate forcings over the past two centuries was possibly as much as +0.5 watts. This would be at least as large as the human-made climate forcing up through the middle of the twentieth century. Therefore natural forcings are a good candidate for explaining, or helping to explain, observed climate change up to the mid-twentieth century.

Climate change between the eighteenth and twentieth centuries was noticeable. The eighteenth century fell within the Little Ice Age, which is sometimes described as taking place from 1600 to 1850 and sometimes from 1250 to 1850. Climate fluctuated from year to year and decade to decade, but changes between the eighteenth and twentieth centuries were significant. In 1780, for example, soldiers in the American Revolution could drag cannons across the frozen New York harbor from Manhattan to Staten Island. The River Thames frequently froze over in the 1700s, and people held ice fairs on it. The twentieth century was too warm for such events to be possible.

Yet how much warmer was the first half of the twentieth century relative to the Little Ice Age? The cooling during the Little Ice Age, averaged over the planet and over the seasons, was probably less than one-half degree Celsius. Our knowledge of both the climate forcing over the past millennium and the climate change are too imprecise to allow empirical evaluation of climate sensitivity.

In contrast, the climate change between modern conditions and the last Major Ice Age, twenty thousand years ago, was an order of magnitude larger. An ice sheet more than a mile thick covered present-day Canada and northern parts of the United States, including Seattle, Minneapolis, and New York City. Another ice sheet covered Europe. The global average temperature was 5 degrees Celsius (9 degrees Fahrenheit) colder than today's. The climate forcing mechanisms that maintained the temperature change were also an order of magnitude larger than the forcings in the last two centuries caused by the sun and volcanoes. The glacial-to-interglacial climate and forcing mechanisms are known well enough to accurately define climate sensitivity (as discussed in chapter 3).

What is clear is that human-made climate forcings added in just the past several decades already dwarf the natural forcings associated with the Little Ice Age. Carbon dioxide increased from 280 parts per million (ppm; thus 0.028 percent of atmospheric molecules) in 1750 to 370 ppm in 2000 (and to 387 ppm in 2009). The impact of this CO_2 change on Earth's radiation balance can be calculated accurately, with an uncertainty of less than 15 percent. The climate forcing due to the 1750–2000 CO_2 increase is about 1.5 watts. Other human-caused changes, such as adding methane, nitrous oxide, chlorofluorocarbons (CFCs), and ozone to the atmosphere, make the total greenhouse gas forcing about 3 watts.

Figure 1 also illustrates a major uncertainty about the *net* human-made climate forcing. The uncertainty is due to aerosols, fine particles in the air that are produced mainly in the burning of fossil fuels. Aerosols scatter and absorb sunlight, reducing the amount that gets to the ground. Sometimes this is called "global dimming." It has a cooling effect that tends to offset greenhouse gas warming, but to an uncertain degree, given the lack of accurate aerosol data. If the estimated climate forcings in figure 1 are accepted at face value, the net climate forcing in 2000 relative to the preindustrial climate was between 1.5 and 2 watts, but with an uncertainty of at least 1 watt.

I had a small 1-watt Christmas tree bulb in my pocket at the Task Force meeting, which I pulled out during my presentation, causing some eyebrows to raise in curiosity. I explained that the net effect of human-made climate forcings was equivalent to having two of those bulbs burning night and day over every square meter of Earth's surface.

I mentioned that in some sense the forcing by two 1-watt bulbs is small—it cannot stop the wind or alter an ongoing weather fluctuation. Yet if it is left in place for decades and centuries, long enough to allow the ocean temperature to fully respond, it is a huge forcing.

Colin Powell, who had returned from his phone call in time for my presentation, asked about the "black soot," or black carbon aerosols—a topic I had hoped would be raised. Black soot, unlike sulfates and other reflective aerosols, absorbs sunlight and thus can cause warming. It is also among the most dangerous of aerosols to human health. Black soot is produced by diesel engines, household coal burning, and stoves that burn field residue or animal waste. It is especially abundant in India, China, and other developing countries.

The point I wanted to make was that a focus on air pollution has practical benefits that unite the interests of developed and developing countries. I had included in my handout a paper that four colleagues and I had published in 2000. We advocated an "alternative scenario" for twenty-first-century climate forcings, alternative to the business-as-usual scenarios studied by the Intergovernmental Panel on Climate Change.

Our alternative scenario emphasized the merits of reducing health-damaging air pollutants, especially black soot, low-level ozone, and methane. We argued that carbon dioxide emissions also would need to be slowly reduced, with atmospheric carbon dioxide kept to a maximum of about 450 ppm (I now realize that carbon dioxide will need to be reduced even further, to 350 ppm, at most). If both of those goals were accomplished, additional global warming, beyond that in 2000, would be less than 1 degree Celsius (1.8 degrees Fahrenheit)—much less than that in IPCC scenarios. Climate effects may still be significant, but they would be far less damaging.

We moved on to Ron's presentation on climate modeling. As Dan had seemed to anticipate, the sledding became difficult as we got into the complexities of climate modeling, the many components that make up the models, and the uncertainties. Some eyes began to glaze over in regard to the complexities of cloud modeling. Cheney interrupted, saying that the topic was important but another session would need to be scheduled to complete it.

Our meeting ended with a curious juxtaposition of comments about the next meeting by O'Neill and Cheney. O'Neill noted that all the present speakers were convinced of the reality of concerns about human-made global warming. He suggested that the next meeting include a global warming contrarian (disbeliever). His rationale was that the Task Force should not be criticized for listening to only one perspective. The rationale might have been sound, but unless there was independent expert arbitration, such as by the National Academy of Sciences, there was a risk that the Task Force would end up simply being perplexed.

Cheney then read statements from the abstract of our alternative scenario paper about the importance of climate forcings other than carbon dioxide. He concluded that the Task Force should hear more from me. The vice president, no doubt, was attracted by our emphasis on less well-known climate forcings, because most of those forcings had sources other than fossil fuels. His specific interest made me uncomfortable, because his choice of speaker seemed to be based on who would deliver an answer that he wanted to hear. That is pretty much the opposite of the scientific method. In science, you want to examine evidence that seems to disagree with your preliminary interpretation. You must evaluate contradictory evidence to make sure that you are not fooling yourself.

If the vice president had read our entire paper carefully, he would have realized that we also called for much less fossil fuel burning than in IPCC scenarios. Nevertheless, being an eternal optimist (what else can be effective?), I welcomed the chance to appear before the Task Force again. My aim would be to clarify that reducing carbon dioxide emissions, as well as air pollutants, was needed to stabilize climate.

As fate had it, three days before the next Task Force meeting, a Chinese fighter jet bumped a U.S. Navy reconnaissance plane, forcing it and its twenty-four crew members to make an emergency landing in China. Because of diplomatic efforts to recover the crew and plane, both the vice president and secretary of state were absent from the second meeting, held at the Environmental Protection Agency headquarters.

They did not miss much.

Richard Lindzen of the Massachusetts Institute of Technology and I were the presenters at the second meeting. Lindzen is the dean of global warming contrarians, the one who is most articulate and has the most impressive scholarly credentials. He was elected to the National Academy of Sciences at a tender age, primarily as a result of brilliant mathematical analyses of atmospheric dynamics.

I knew what I would be up against. In a situation like this, presentation and style are as important as substance. Lindzen is soft-spoken but has an authoritative air; he never loses his cool and is always in complete control. He and other contrarians tend to act like lawyers defending a client, in my opinion, presenting only arguments that favor their client. This is in direct contradiction to my favorite description of the scientific method, by Richard Feynman: "The only way to have real success in science . . . is to describe the evidence very carefully without regard to the way you feel it should be. If you have a theory, you must try to explain what's good about it and what's bad about it equally. In science you learn a kind of standard integrity and honesty."

The scientific method, in one sense, is a handicap in a debate before a nonscientist audience. It works great for advancing knowledge, but to the public it can seem wishy-washy and confounding: "on the one hand, this; on the other hand, that." The difference between scientist-style and lawyer-style tends to favor the contrarian in a discussion before an audience that is not expert in the science.

I long ago realized that the global warming "debate," in the public mind, would be long-running. I also noted that contrarians kept changing their arguments as the real-world evidence for global warming continued to strengthen, conveniently forgetting prior statements that were proven wrong. For that reason, when I publicly debated Lindzen in 1998, and contrarian Pat Michaels a few weeks later, I decided that the best approach was to make a table of our basic differences.

I knew that I could not "win" the debates, which inevitably appear to the public to be a technical dispute between theorists. Instead, I wanted to pin down our differences so that some years in the future a thoughtful person could make an objective assessment. That table was published in *Social Epistemology* in 2000 and is reproduced in appendix 1 (page 279).

My plan for the second Task Force meeting was to first summarize my conclusions from the first meeting and discuss the relevance of the information for policy. Then I would deal with the contrarian perspective with a single chart, my "table of differences." I brought a transparency of that table (a viewgraph—I was a holdout from Power-Point for many years).

I started by again showing the climate forcings bar graph (figure 1) and drawing big circles around the bars representing methane (CH_4), ozone, and black carbon aerosols (black soot). I noted that together these health-damaging air pollutants contributed as much climate forcing as carbon dioxide.

Then I showed a chart (discussed in chapter 3) providing evidence from Earth's history for how sensitive global climate is to a change of climate forcings. The chief implication is that additional human-made climate forcing, above that in the year 2000, should be kept to less than 1 watt. If we succeed in that, further global warming should not exceed 1 degree Celsius. However, if added climate forcings exceed 1 watt, global temperature would be pushed well above the range that has existed for the past million years. Global warming of 2 degrees Celsius or more would make Earth as warm as it had been in the Pliocene, three million years ago. Pliocene warmth caused sea levels to be about twenty-five meters (eighty feet) higher than they are today.

I concluded this discussion with a diagram that contrasted the business-as-usual scenarios that IPCC examined and my alternative scenario, which defined a course that would keep additional global warming at less than 1 degree Celsius, thus presumably allowing Earth to retain a climate resembling that in which civilization developed.

The business-as-usual scenario increased the carbon dioxide forcing 2 watts between 2000 and 2050, and increased the non–carbon dioxide forcings 1 watt, for a total of 3 watts. The alternative scenario, in contrast, required a slow reduction of carbon dioxide emissions over fifty years, keeping added carbon dioxide forcing at 1 watt. Net non–carbon dioxide forcing is kept at zero by reducing black soot, ozone, and methane enough to counter expected added climate forcing due to increasing nitrous oxide (from use of fertilizers) and decreasing sulfate aerosols (from cleaning up air pollution).

Achievement of the alternative scenario would require two major policy actions. First, a downward trend in carbon dioxide emissions requires an increase in energy efficiency and the use of renewable or other energies that do not produce carbon dioxide. Second, reduction of the non–carbon dioxide climate forcings requires global programs to reduce air pollutants that contribute to global warming (black soot, ozone, and methane). I reiterated that the combination of these two policy goals could unite the interests of developed and developing countries.

Finally, I showed the table contrasting my position and that of Richard Lindzen. The first item concerned Lindzen's take on the magnitude of global warming when it became a public issue in the late 1980s. He had stated repeatedly, consistent with an analysis by his MIT colleague Reggie Newell that was later shown to be flawed, that global warming over the prior century was only about 0.1 degree Celsius. In contrast, I had reported in the late 1980s that global warming was about 0.6 degree Celsius. Numerous later studies had confirmed my conclusion, and subsequent additional warming had increased total global warming to almost 0.8 degree Celsius.

But Lindzen was prepared. Before I could move to the second point, he interjected in his calm, unflappable style, "The reference you have given, *MIT Tech Talk*, is basically a newspaper." Turning to the cabinet members, he added, "You all know how accurate newspaper quotes are." There were a lot of nods and chuckles. I was aware that Lindzen had made his assertion about the near absence of global warming many times in the late 1980s—it was not a misquote by a writer—yet before this audience, Lindzen had won the point.

A similar problem, to a lesser degree, occurred with regard to other items in appendix 1. Scientists who have heard Lindzen are well aware of his statements about observed global warming, climate sensitivity, water vapor feedback, and so forth. If the positions Lindzen had expressed on these matters went before, say, the National Academy of Sciences, he obviously would be on the defensive. But on what basis are cabinet members going to choose between academics with opposing views?

The capable lawyer knows that oral statements can be dismissed as hearsay. My failure to adequately appreciate this for several years caused other problems, as I relate in later chapters.

Lindzen used part of his presentation to show graphs of observed data such as temperature and precipitation, emphasizing the large fluctuations and possible measurement errors. His aim seemed to be a conclusion that global warming is a very uncertain proposition. He focused on more local observations, because he could no longer dispute the reality of global warming. But he had managed to defuse his earlier assertion about the absence of global warming, which had been proven to be wrong.

Lindzen also spent substantial time questioning the motives of scientists who, he said, made "alarmist" statements. His thesis was that most scientists concurred with the reality of global warming only because it increased their ability to obtain research funding. If I had been on my toes, I could have pointed out that in 1981 I had lost funding for research on the climate effects of carbon dioxide because the Energy Department was displeased with a paper, "Climate Impact of Increasing Atmospheric Carbon Dioxide," I had published in *Science* magazine. The paper made a number of predictions for the twenty-first century, including "opening of the fabled Northwest Passage," which the Energy Department considered to be alarmist but which have since proven to be accurate.

Unfortunately, this second meeting served to confuse Task Force members, rather than illuminate them. I heard indirectly from presenters at subsequent meetings that Task Force members had related that they could not evaluate our contrasting viewpoints. The third Task Force meeting focused on economics and included Richard Schmalensee of MIT as a presenter. The fourth meeting focused on policy, including the Kyoto Protocol, and had former EPA administrator Bill Reilly and Kevin Fay, a representative of the business community, as presenters. Additional Task Force meetings may have been held prior to the president's June 11, 2001, Rose Garden speech summarizing the administration's climate and energy policies (discussed in chapter 3).

After the second Task Force meeting, I shared a taxi with Richard Lindzen. We had always been cordial with each other, but not much was said during this ride. I was feeling down, realizing that my optimism at the end of the first meeting had been a mistake. A draw in a global warming "debate" is a loss, because policy inaction is the aim of those who dispute global warming. As we pulled up alongside a Chrysler PT Cruiser, I broke the silence by commenting that it seemed to be an interesting throwback. He said that it was cute but did not have enough trunk space.

I considered asking Lindzen if he still believed there was no connection between smoking and lung cancer. He had been a witness for tobacco companies decades earlier, questioning the reliability of statistical connections between smoking and health problems. But I decided that would be too confrontational. When I met him at a later conference, I did ask that question, and was surprised by his response: He began rattling off all the problems with the data relating smoking to health problems, which was closely analogous to his views of climate data.

Oof, I thought—if I had asked him about the relation between smoking and cancer during the Task Force meeting, his response might have been revealing, almost like Jack Nicholson's "You can't handle the truth!" in *A Few Good Men*. Or maybe not. It is not likely to be that easy.

CHAPTER 2

The A-Team and the Secretary's Quandary

ON THE WAY HOME FROM THE SECOND Climate Task Force meeting, I had an idea. In my bag was a reminder about an overdue assignment: I needed to define the next project for my student-teacher research team. My idea was to have the team work on a project that would help clarify what I had failed to do a good job of explaining during the Task Force meetings—the implications of climate change for energy use.

Several years earlier, my colleague Carolyn Harris and I had initiated a research education program that we called the Institute on Climate and Planets. Each summer we would work with students, teachers, and professors from several New York City high schools, ranging from the disadvantaged to the highly competitive Bronx High School of Science, and from a few colleges in the City University of New York system, ranging from two-year community colleges to the City College of New York. The program had simultaneous objectives in science education, research experience, and minority participation.

Participants were divided into several teams. My team typically had about ten people, including two high school students, two high school teachers, two or three college students, a college professor, myself, and sometimes one or two other scientists. I would define a research problem, and we would work on it as a team over the summer, with some students continuing to work on it through the academic year. The educators used their experience with the research problem in their science classes and in special after-hours research courses at their schools.

For the first two years of this program, which started in the mid-1990s, my team was called Pinatubo. Our aim was to use the 1991 eruption of Mount Pinatubo, the largest volcanic eruption in the twentieth century, as a natural climate experiment, helping us to understand climate processes. One of our tools was a global climate model, which I had helped develop over the previous two decades. We used the model in combination with global climate observations, making climate simulations and examining how well the model could reproduce the climate variations following the volcanic eruption. At the end of our research, in 1996, we published a paper, "A Pinatubo Climate Modeling Investigation," in the book *Global Environmental Change*.

My next team, with some new students, was called Forcings and Chaos, and for a few years we compared climate simulations covering two decades with observations in an attempt to disentangle climate change driven by forcings from unforced chaotic climate variability. Our work was published in the *Journal of Geophysical Research* in 1997.

I renamed my group the A-Team—from "alternative scenario"—for the new project. Their assignment: To imagine the secretary of state needs the A-Team's help in devising a strategy to deal with global warming. The team must analyze climate and energy data and report back to the secretary, so he can advise the president on what actions need to be taken to save the planet. (I assumed that Colin Powell—and Paul O'Neill—must have been puzzled by Energy Secretary Abraham's claim that the United States needed to build ninety new coal-fired power plants every year for the next twenty years.) I titled the student's task description for this imaginary scenario "The Secretary's Quandary." It started like this:

THE SECRETARY'S QUANDARY

The secretary of state is caught between a rock and a hard place. As leader of the State Department, he deals with countries around the world. These countries are calling for the United States to reduce its emissions of CO_2, the principal gas that stands accused of bringing on dreaded global warming.

Different perspectives. Yet the secretary knows that the Department of Energy has a different perspective. Its job is to assure

that the United States has a supply of affordable energy sufficient to drive a strong economy. All parties, the president and
his cabinet, agree that a strong economy is needed to produce
the technology development and the resources required to eventually stabilize atmospheric composition and solve the global
warming problem.

It is also realized that the long-term solution of global warming will require many decades. Fossil fuels (coal, oil, and gas)
produce CO_2, these fuels power our economy, and the lifetime
of energy infrastructure can be many decades. A strategy to deal
with global warming must be devised in concert with continuing technology development and improvements in understanding of climate science.

The big issue concerns actions that could be taken now to
slow the growth of CO_2. Recent research has shown that if the
growth rate of CO_2 emissions could be stabilized and then begin to decline, climate change would be moderate—some global
warming would be expected, but the danger of disastrous climate
change would be much reduced. Prompt leveling of CO_2 emission rates would provide time to develop improved technologies
and an economically sound strategy to reduce CO_2 emissions
and stabilize climate.

The official bottom line. The secretary of state is troubled
because the Energy Department has advised the president that
the United States cannot stabilize its CO_2 emissions in the
next decade. To be sure, dedicated Energy Department employees have made great strides in advancing the potential of energy
efficiencies in homes and in industry. Yet the official bottom
line is that, even with improved energy efficiencies, CO_2 emissions will need to increase 15 percent in the next decade to provide healthy economic growth.

The secretary realizes that, as he travels around the world with
this energy plan, he will be severely beaten about the head and
shoulders, at least in a figurative sense, in many countries. His
disquiet arises, however, because he has come to realize that the
climate change issue has at least some validity, and with this energy plan the United States will aggravate future climate problems for the young and the unborn.

Besides, the secretary has a nagging feeling that something is inconsistent in the energy and CO_2 projections. He knows the growth rate of energy use and CO_2 emissions in the United States has been moderate in the past three decades, only about 1 percent per year, as opposed to 4 percent per year in the previous century. He also knows the president has publicly favored aggressive new actions to improve energy efficiency and develop renewable energies. Yet official projections have energy use and carbon dioxide emissions increasing at a rate at least as large as in recent decades. Something doesn't square up.

A team player's quandary. The secretary's quandary arises because he knows that the president must rely on his Energy Department for projecting energy needs, and the secretary is a consummate team player. What can he do? His first thought is to fiddle with the energy and CO_2 numbers himself, and to try to figure out if something is wrong. After all, like Benjamin Franklin, the secretary is a bit of an amateur scientist (well, not quite like Benjamin Franklin). But he soon realizes the futility. He is dealing every day with crises in the Middle East and around the world, including the war on terror.

Suddenly, an idea hits him—he must call on the A-Team, a group of students and teachers he knows in the New York City area. He and the president are committed to young people and their education. Who better to investigate this problem than the people who will inherit the consequences of our energy plan?

The scientific approach. The A-Team enters. They look ragtag—some bleary-eyed students, a couple of energetic teachers, a wizened professor—but the secretary doesn't mind their appearance. He knows that they take a scientific approach; they give primacy to real data. Theories and models of the future can help organize one's thoughts, but they are only useful if they explain the real world. A convincing analysis must start with and place most weight on data and real-world observations.

Their job, the secretary explains, is to provide a hard-nosed analysis, one that can be taken to the president to help him. The president is besieged from both sides. Environmental advocates see the world through their lenses—they are not concerned about the health of industry. And energy advocates argue that we must have more and more energy—climate change may be exaggerated, they say, and future generations can deal with it. The president's job is tough, and he needs some objective scientific help.

One good graph. The secretary can provide the A-Team with only one graph. "One good graph is worth a million words," the secretary says. Staring at the chart (figure 2), he says, "This graph defines the enigma. Perhaps it can also help you define your analysis of the problem."

The scenarios. The graph contrasts two energy paths for the United States that were proposed in the mid-1970s. The Energy Department projected the need for strong energy growth rates. It said that U.S. energy consumption of 70 quadrillion BTU annually in 1975 would need to increase to 200 quadrillion BTU annually forty years later, a growth rate of about 3 percent per year.

An extreme alternative to the Energy Department scenario was provided by Amory B. Lovins, an idealist and a renowned visionary. His scenario has continual improvements in energy efficiency, so energy use grows only slightly and then begins to decline. In addition, more and more of that energy is produced by what Lovins describes as "soft technologies," ones that do

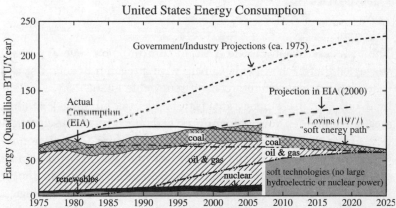

FIGURE 2. *U.S. energy consumption falls well below government and industry projections, even below projections made by the Department of Energy's Energy Information Administration (EIA) in 2000. However, Amory B. Lovins's projection (in* Soft Energy Paths: Toward a Durable Peace, *Penguin Books, 1977) that fossil fuels, nuclear power, and large hydroelectric power would all be largely replaced by small-scale renewable energy has also proved to be inaccurate.*

not include nuclear power or big hydroelectric plants—energy sources that are also the banes of some environmentalists.

CO_2 emissions (from coal, oil, and gas) in Lovins's scenario decline dramatically, almost disappearing by 2025. The students noted that his scenario is more extreme than the "alternative scenario." CO_2 emissions in their "A-scenario" peak early in the twenty-first century and decline enough by midcentury to prevent global warming from exceeding 1 degree Celsius. The students are puzzled because their A-scenario is already much more ambitious than those considered by the Intergovernmental Panel on Climate Change. So they are eager to see how Lovins's even more optimistic scenario compares with the real world.

The real world. The real-world data for energy use in the United States (the EIA curve in figure 2) show that Lovins was at least half right. U.S. energy use grew only slowly, about 1 percent per year, after 1975. But the data also show that Lovins's scenario, if taken as a prediction, was half wrong, at least so far. Use of renewable energies such as the sun and wind is still so small that it barely shows up in the graph. Yet the A-Team could not dismiss Lovins as a dreamer—perhaps energy policies ignored opportunities, and Lovins was just ahead of his time by a few decades.

The task description for the A-Team continued for several pages. I suggested that each student pair up with a teacher or scientist and that each pair choose one form of renewable energy (say, geothermal), one energy efficiency area (say, residential buildings), and one area of technology development (say, carbon capture and sequestration). Then they would estimate the potential for CO_2 emission reductions for fifteen-year and thirty-year time horizons—the shorter period would need to rely on existing technology, while the longer period could include realistic projections for improved technologies. They also would estimate results with "current trends" (no policy changes), "moderate action" (actions with little or no cost), and "strong action" (government-mandated energy reforms or technology subsidies).

The A-Team was the most enthusiastic and hardworking of all my Institute on Climate and Planets teams. The pairs reported back to the full team on a weekly basis, and by the end of the summer they had made good progress on the renewable energy and energy efficiency tasks. I wrote a letter to Colin Powell inviting him to give the keynote speech at our summer institute closing ceremony; we wanted to also show him the A-Team's results. Unfortunately, Secretary Powell could not attend, but this did not deter the A-Team from continuing their enthusiastic work for two more years.

We decided to go into greatest detail on automobile efficiencies during the next two years, for several reasons. Vehicles were the fastest-growing source of CO_2 emissions, yet the government seemed to be oblivious to the matter. Also, during the course of the A-Team study, I had been invited to give a talk at Exxon/Mobil headquarters, and my discussion of this experience with the A-Team spurred our desire to focus on vehicles.

My talk at Exxon/Mobil was to executives and top engineers of the ten leading automobile manufacturers in the United States, including Japanese and European companies. Afterward, we had a friendly question-and-answer session that addressed climate model uncertainties, causes of climate change in Earth's history, and my assertion that we must get onto a path resembling the alternative scenario in order to avoid disastrous climate change.

I stayed for the rest of the morning as engineers described their plans. Criticisms of the California Air Resources Board (CARB), which was attempting to force the car manufacturers to deliver major improvements in vehicle efficiency, grew more and more strident. Finally I raised my hand and asked, "Wouldn't it make sense, instead of fighting CARB, to try to get ahead of the curve by focusing on vehicle efficiency?" The response was, "Dr. Hansen, we have to give the customers what they will buy, and they want higher performance and larger vehicles."

That evening I noticed several television advertisements showing huge vehicles parked atop mountain peaks (where probably nobody would ever actually drive). This led me to question how much of the desire for size and performance really originated with the customers.

Nevertheless, the A-Team decided to look at it this way: For twenty years the automotive industry has used advances in technology only to increase vehicle size and performance, keeping average miles per gallon at about the same level. Vehicles now had size and performance. If technology gains in the *next* fifteen years were used to improve efficiency, retaining current performance levels, how much could miles per gallon be increased?

Relying heavily on a recent National Research Council (NRC) study of potential vehicle-efficiency improvements, the A-Team wrote a report considering different degrees of technology infusion, ranging from changes that the NRC deemed "production ready" to emerging technologies.

The A-Team also developed the Auto CO_2 Tool and made it available on the Web. Users could make alternative assumptions for technology infusion and view graphs of the results, including reductions in national oil requirements. The Auto CO_2 Tool showed that moderately aggressive improvements in efficiency resulted in efficiency gains ranging from 18 percent for subcompact autos to 37 percent for large pickups. Vehicle price increases, ranging from eight hundred to three thousand dollars, generally were covered by reduced fuel costs within a few years.

The Auto CO_2 Tool revealed that even without further vehicle-efficiency improvements, if the existing production-ready vehicle-efficiency improvements were made, the resulting reduction in U.S. oil imports by 2050 would be 7 ANWRs, where 1 ANWR is the entire amount of oil that the United States Geological Survey estimates to be recoverable from the Arctic National Wildlife Refuge. Once the vehicles with these moderately aggressive fuel-efficiency improvements were fully phased into the vehicle fleet, which requires about fifteen years, the annual reduction in U.S. oil import costs, with oil at fifty dollars a barrel, would be a hundred billion dollars.

The A-Team was counting on me to write a publishable paper about the results of our study, which, in my opinion, were significant. Our basic conclusion was that existing technology could provide a CO_2 emissions path consistent with the alternative scenario in the near term, i.e., a downturn of emissions over the next fifteen years, as opposed to the continual emissions growth that the Energy Department projected. This conclusion applied not only to the transportation sector but also to the industrial, commercial, and residential sectors.

The paper I wrote, "On the Road to Climate Stability: The Parable of the Secretary," took a novel approach. It included twenty-two graphs and pie charts organized into eight figures, but the text of the paper was written partly as a discussion among three A-Team members: the wizened professor (some combination of York College professor Sam Borenstein and me) and two students, Jorge and Naomi (representing all the other A-Team members). I aimed to provide a more complete picture of the research process by capturing the emotions as well as scientific results.

Unfortunately, journal editors and referees did not like this approach, and my attempts to get the paper published were rebuffed. We lost funding for the education program, and efforts to get foundation funding failed. Carolyn Harris moved back to Washington, where she had once been a high school teacher, and I had a full-time regular job and was writing several other papers (especially "Can We Defuse the Global Warming Time Bomb?"), so the A-Team paper never had the audience it deserved.

Below is the final section of the paper. In the preceding section the A-Team had just completed its report, and the students were musing about whether the nation's leaders would have the gumption to take the actions necessary to achieve the potential energy efficiencies the team had found, including battling special interests that defend the status quo.

The wizened professor said to them, "But aren't we getting out of our area of knowledge? We did our best. Let's give the report and the Auto CO_2 Tool to the secretary. He can take it to the Task Force on Energy and Climate."

The secretary's debriefing, of course, came from our imagination.

DEBRIEFING BY THE SECRETARY

The A-Team waited nervously for the secretary to return from a meeting of the Task Force on Energy and Climate. What could be taking so long?

When they finally saw him coming down the corridor, their faces fell, as he bore an uncharacteristic and distant scowl. Seeing them, he brightened and said, "Why so glum? The Task Force members were enthusiastic about your report and agreed with the thrust of it. They were impressed by your work."

"The chairman?"

"The chairman liked it too. He agreed with most things that

you wrote. He will adopt much of the language in future positions and official statements. He sends his sincere thanks for your efforts."

"You mean that actions will be taken? Does he expect to meet the CO_2 emissions in the A-scenario? How will it be done?"

"I am afraid that will not be in the plans. He agrees with the need for technology development. Maybe in the future it will be possible to slow emissions."

"But you said he agreed with our analysis and would adopt the language in official statements. We showed that it was practical and beneficial to reduce emissions with existing technology, no?"

"Well, he did not entirely agree with the report. He did not think that energy needs would be slowed or that CO_2 emissions could be reduced."

"But didn't we show that it was possible with existing technology? What did he say about that part of the report?"

The secretary paused. "He said that part of the report was . . . naïve."

"Well," the professor interjected, "there are other important components to energy use and CO_2 emissions. What about energy efficiencies in households and buildings?"

"The Task Force agrees there is great potential for savings in the United States. It is up to Congress and the states to lead. There will be an effort to cooperate with Congress, especially on gasohol."

"Gasohol?"

"Yes. Didn't you agree that it could be done in a way that would be marginally useful?"

"We didn't spend much time on it," Jorge said. "We looked into it enough to conclude that the assumptions of the Berkeley professor who had argued that gasohol took more energy to produce than it provided were not necessarily right."

"However," said the professor, "gasohol is surely a minor player in the overall energy and CO_2 problems. Why give it priority?"

The secretary looked away and didn't answer. He seemed to be musing about the larger picture.

A long period of silence was broken by the professor. "What are you going to do now?"

"I don't know," the secretary said thoughtfully. "We are all team players."

Naomi ended another period of silence by asking, "Do you think we are naïve?"

The secretary hesitated, but he had regained his composure. "*Naïve* is an interesting word. It has more than one connotation. Perhaps we are all naïve."

CHAPTER 3

A Visit to the White House

My presentations to the Climate Task Force had been ineffectual. Our A-Team results helped make clear that the energy policies needed to safeguard climate would also be in the best interests of the nation, but I had failed to get that work published in a science journal.

Fortunately, in June 2003 I received another opportunity to communicate: an invitation to give a presentation to the most effective levels at the White House. The invitation was from Jim Connaughton, chairman of the White House Council on Environmental Quality (CEQ), who was widely recognized as the most powerful person in the Bush-Cheney administration on climate matters, aside from, of course, the president and vice president.

I am not sure why I received the invitation in 2003. Perhaps it was again because of my published statements about the importance on non–carbon dioxide climate forcings, which the vice president had read at the first Task Force meeting.

I had the impression that the administration respected me but was leery. For example, I was invited to go with the U.S. delegation to the G8 Environmental Futures Forum in Spoleto, Italy, in October 2001. However, I had a special status: I was with the U.S. delegation, but I was not officially part of it. That way, if I said anything that was not appreciated by the administration, it would not be misinterpreted as a position of our government. I realized that I was considered to be a maverick. Everyone was aware that I had complained publicly about the White House's alteration of my testimony to Al Gore's Senate committee during the first George

H. W. Bush administration. My complaint had caused a good deal of consternation, and John Sununu, Bush's chief of staff, attempted to have me fired. I was able to keep my job because John Heinz, Republican senator from Pennsylvania, intervened on my behalf. All seemed to be forgiven now, but wariness remained.

I should note again here my political inclinations. I am a registered Independent. I believe that the United States would benefit from a third party. (I did not vote for the Texan who saw Martians in his front yard, but I probably would vote for someone like Mayor Michael Bloomberg of New York.) Our biggest problem, in my opinion, is due to the role of money in government, the special interests, epitomized by hordes of lobbyists in expensive alligator shoes. The issues that most influence my preference in political candidates are campaign finance reform and environmental and climate change policies. I supported the Gore-Lieberman campaign, to which I contributed a thousand dollars.

Yet early in the Bush-Cheney administration I was hopeful of a turn toward more effective actions regarding climate change. The Clinton-Gore administration had been ineffectual in this matter—and that is being generous; emissions actually increased substantially under their leadership. I thought that Bush would surely oppose the Kyoto Protocol, which deserved many of its criticisms, but that he might be able to work effectively with the business community and Congress if he chose to take actions. Also, I had the impression that Ari Patrinos might be listened to. Patrinos was an outstanding scientist at the Department of Energy who had supported research on climate change. From exchanges with him, I knew that he was helping with position statements related to climate.

Of course, by June 2003 it was clear that Bush would pull no "Nixon goes to China" act with regard to climate change. But I had not yet been openly critical of the administration, so maybe I would be listened to in this trip to the White House. On this visit, I would have enough time for a clearer presentation, and I unambiguously titled my talk "Can We Defuse the Global Warming Time Bomb?" But before discussing my talk, I need to describe the situation.

The U.S. policy on global warming in June 2003 remained as it was in June 2001, when President Bush had given a well-prepared Rose Garden speech defining the country's position on the subject in detail, including many valid statements. The vice president and Climate Task Force members stood with Bush as he delivered the speech, which bore Patrinos's imprint. The president noted: "My cabinet-level working group has met regularly for the last ten weeks to review the most recent, most accurate, and most comprehensive science."

The impression the speech created, aided by a widely distributed photo of Bush closely surrounded by Task Force members in the Rose Garden, was misleading in one critical sense. The most important policy position that the president promulgated was not arrived at on the basis of the Climate Task Force meetings, and science had little to do with that key decision.

Recall that George W. Bush came into office carrying a pledge to treat carbon dioxide as a pollutant. When EPA administrator Christine Todd Whitman testified on February 27, 2001, to a Senate Committee on Environment and Public Works subcommittee, she advocated a plan for regulating carbon dioxide emissions under the Clean Air Act. At an international meeting the following week, Whitman said that she "assured [her] G8 counterparts that the president's campaign commitment to seek a mandatory cap on carbon dioxide emissions was solid," according to her book, *It's My Party, Too*.

The promise remained until her words spurred actions behind the scenes, culminating in an infamous March 6 letter to President Bush from senators Chuck Hagel, Jesse Helms, Pat Roberts, and Larry Craig. That letter drew attention to Whitman's remarks and asked the president to clarify the "Administration's position on climate change, in particular the Kyoto Protocol, and the regulation of carbon dioxide under the Clean Air Act."

Bush responded with a March 13 letter to Senator Hagel in which the president reversed his position on carbon dioxide, stating that it was not a pollutant under the Clean Air Act. He claimed that important new information warranted the reevaluation, specifically a Department of Energy report concluding that caps on carbon dioxide emissions would reduce the use of coal and raise the price of electricity.

Analysis of the March 6 and March 13 letters, and of what happened behind the scenes, is contained in Mark Bowen's *Censoring Science*, Whitman's *It's My Party, Too*, and *The Price of Loyalty* by

Ron Suskind, with the cooperation of Paul O'Neill. O'Neill notes that the tone and much of the substance of the letter from the four senators seemed to have come "right out of Dick Cheney's mouth" and that he believes the letter from the president in response was prepared by the vice president.

Suskind describes Cheney as a puppeteer pulling strings. According to Suskind, Whitman went to the Oval Office on the morning of March 13 hoping to argue her case, but instead Bush read to her portions of the letter reneging on his pledge to regulate carbon dioxide. As Whitman left the Oval Office, Cheney arrived expressly to pick up the letter—which he pocketed and took to his weekly policy meeting with Republican senators.

Whether this turn of events was, as Suskind describes, the "clean kill" of a puppeteer is not the important point. Rather, as Bowen puts it, the episode was a "knockout punch to facts-based consensus-building decision-making." The decision not to regulate carbon dioxide had been made two weeks before the first meeting of the Climate Task Force, which was supposed to consider the evidence.

Moreover, to the extent that there was any mention of science in the March 6 and March 13 letters, it involved a faulty interpretation of our "alternative scenario" paper. The four senators stated in their letter: "In August 2000, Dr. Hansen issued a new analysis which said the emphasis on carbon dioxide may be misplaced. In his new report, he stated that other greenhouse gases—such as methane, black soot, CFCs, and the compounds that create smog—may be causing more damage than carbon dioxide and efforts to affect climate change should focus on these other gases. 'The prospects for having a modest climate impact instead of a disaster are quite good, I think,' Dr. Hansen was quoted as saying in the *New York Times*."

In retrospect, I had made at least two mistakes. The first was my wording in the alternative scenario paper. I aimed to draw attention to the importance of non–carbon dioxide climate forcings, but only to give them their proper due, not to allow an escape hatch for carbon dioxide. My alternative scenario required, in addition to absolute reductions of the non-CO_2 forcings, aggressive efforts to slow the growth of carbon dioxide emissions. Specifically, the annual growth of atmospheric carbon dioxide, which was averaging 1.7 ppm per year at the end of the twentieth century, would need to slow to 1.3 ppm per year by 2050 in order to achieve the alternative scenario. If this happened, additional climate forcing would be limited to about 1 watt and additional warming, after 2000, would be less than 1 degree Celsius. However, as mentioned earlier, to achieve such a slowdown in the growth rate would require a strong emphasis on energy efficiency, as well as a steady increase in the use of renewable energies or other energy sources, such as nuclear power, that produce little or no carbon dioxide.

In contrast, on March 19, ten days before the first Climate Task Force meeting, President Bush and Energy Secretary Spencer Abraham discussed with the media the need to increase the supplies and the use of fossil fuels. Abraham specifically mentioned plans to open up the Arctic National Wildlife Refuge to oil and gas drilling and coal mining, and generally noted the need to open up federal lands and offshore regions to such drilling and mining. As mentioned, Abraham argued that many new coal-fired power plants would be needed.

My second mistake was my failure to emphatically state to the Task Force that the administration's energy plans, as described by Abraham, were in dramatic conflict with the alternative scenario. That conclusion should have been obvious from my presentations to the Task Force as well as from a reading of the alternative scenario paper. But did Task Force members pay sufficient attention or read the handouts we provided? If I had made an explicit, unequivocal statement to start off my presentation at the first Task Force meeting, I may have gotten their attention and perhaps provoked more useful discussion. But then, I probably would not have been invited back—which wouldn't have been such a loss, as the second Task Force meeting, the one including Richard Lindzen, had not been fruitful.

My alternative scenario paper, with its emphasis on the importance of non–carbon dioxide climate forcings, was controversial immediately upon its publication in 2000, in part because environmentalists recognized the possibility that it could be misused by those who preferred there be no restrictions on carbon dioxide emissions. This matter warrants discussion, because it relates to the topic of how science research should be communicated to the public and policy makers, and so I return to it at the beginning of chapter 5. For the moment, though, I want to note that my emphasis on the importance of the various other climate forcings was likely the reason that I was invited to the White House in 2003.

In addition to CEQ chairman Jim Connaughton, presidential science adviser Jack Marburger was expected to attend my White House presentation, along with significant officials from the Office of Management and Budget (OMB) and other members of the administration.

This group would likely be a more significant one than the Climate Task Force as far as implementing specific government actions. The vice president and cabinet members might have been top policy makers, but OMB and CEQ are where things happen. They receive guidance from above but have flexibility in deciding what actually gets done. The highest levels of government usually have little time or inclination to interfere with technical decisions.

Connaughton and CEQ were known to have the day-to-day power within the administration on global warming and climate change matters. Marburger, although he had the title of White House science adviser, and had visited my laboratory in 2002 to hear about global warming, may have been either ineffectual or uninterested in the topic—or perhaps was not as well trusted by the vice president.

In *Censoring Science*, Bowen writes this passage about Marburger and Connaughton, in which "Jim" refers to me: "According to Jim, physicist John Marburger, the director of the Office of Science and Technology Policy and the official science adviser to the president, has nowhere near Connaughton's power—nor his Machiavellian intent." I am pretty sure the last four words in that passage are probably Bowen's words and interpretation. I do not remember details of the two-day interview that I gave to him before he started to write his book, but I would be surprised if I said anything resembling those last four words. That would not square with my impression of Jim Connaughton in 2003, which was positive.

On the other hand, Mark Bowen does a remarkable job of documenting, in detail, the activities of Connaughton and his staff, especially Connaughton's chief of staff, Philip Cooney. Bowen's evaluations are based on documents, including e-mails and letters, some obtained via the Freedom of Information Act. They include exchanges between Cooney and Kevin O'Donovan, the vice president's special assistant for domestic policy, described as Cheney's point man on climate, as well as exchanges among Cooney, Connaughton, and the president's political adviser Karl Rove.

Bowen makes clear that editing and censorship of science that cannot be rationalized under reasonable scientific standards was carried out by Cooney and others under Connaughton's direction and approval. Bowen also concludes that one of the phone calls from the White House to NASA headquarters in December 2005, which ignited a so-called "shitstorm" and attempts to isolate me from the media (events discussed in chapter 7), was probably from Connaughton.

Can these disparate perceptions of Connaughton be squared up? I will try to do so, but I should first describe my presentation to CEQ and its reception.

The theme of my White House presentation (an abbreviated version of which was eventually published in the March 2004 *Scientific American* under the title "Defusing the Global Warming Time Bomb") was that paleoclimate information provides precise knowledge of how sensitive climate is to changes of climate forcings. Human-made forcings are beginning to warm the world at a predicted rate. The limit on permitted global warming, if we wish to preserve the great ice sheets on Antarctica and Greenland, and thus preserve the coastlines that have existed for the past seven thousand years, is much less than has generally been assumed. Halting global warming is still feasible—but requires international cooperation in taking urgent, unprecedented actions, which would have additional benefits for human health, agriculture, and the environment.

It was a good, friendly discussion at the White House, with Connaughton and about twenty-five others, including representatives from OMB and the Office of Science and Technology Policy (OSTP). Marburger was not there—he was on a plane, it was explained—which was just as well, as his deputy, Kathie Olsen, was present. Olsen, more engaged and engaging than Marburger, had been chief scientist at NASA headquarters from 1999 to 2002.

In the beginning of my talk I said that there was bad news and good news about global climate. The bad news: It has become clear that Earth's climate is very sensitive to climate forcings, and we are close to driving the system into a region with dangerous consequences for humanity.

The good news: It is not too late to solve the problem, and there would be multiple side benefits to doing so. I noted that the United States, under George Bush the elder, helped bring about the United Nations Framework Convention on Climate Change, which has the specific objective of stabilizing atmospheric greenhouse gases at a level that avoids dangerous human-made interference with climate. I argued that the actions required to stabilize climate, which require addressing both non–carbon dioxide and carbon dioxide emissions, would likely have economic benefits and would certainly be beneficial for energy security and national security.

During my talk, I noted a quote from the president's June 2001 Rose Garden speech: "We will be guided by several basic principles. Our approach must be consistent with the long-term goal of stabilizing greenhouse gas concentrations in the atmosphere. Our actions should be measured as we learn more from science and build on it. Our approach must be flexible to adjust to new information . . . We will act, learn, and act again, adjusting our approaches as science advances and technology evolves." I suspected these words had flowed from the pen of Ari Patrinos, but they fit my talk perfectly, because, I argued, I now had improved information about the "global warming time bomb" threat and the actions needed to avert it.

The most important scientific insight that I hoped to convey at the White House meeting was based on Earth's climate history, that is, paleoclimate. Climate history is our best source of information about how sensitive the climate system is, and, it turns out, the climate is remarkably sensitive—large climate changes can occur in response to even small forcings.

I mentioned in the preface to this book that understanding climate forcings, imposed perturbations of the planet's energy balance, would be the most difficult science you would need to deal with. Sorry. I was leading you on a bit, hoping to get your toes wet.

Paleoclimate and climate sensitivity might give you more trouble—perhaps not; it is pretty simple. But you are facing a case of double or nothing—you need to decide whether you are willing to learn about them, or whether you prefer not to bother.

If you prefer to remain in the land of the blissfully ignorant, you will have lots of company. Even some scientists, seeing Al Gore mount an elevator contraption and point to paleoclimate carbon dioxide and temperature records in the movie *An Inconvenient Truth*, assert: "He has the science all wrong!" Actually, Gore understands the science well enough, and he had the implications right—he just failed to explain the science.

But if you are willing to expend a modicum of effort, you can take a big step toward appreciating the degree to which we are living on a planet in peril. Additional steps will be needed, but this first one—learning about climate sensitivity and paleoclimate—is essential to developing a realistic understanding of the potential implications of climate change for your children and grandchildren.

All right, it's true that the initiation fee—the "modicum of effort"—is not necessarily small, depending on your training and the time you have to concentrate. If you prefer not to pay these dues, at least not at the moment, perhaps rather than casting the book aside, you might skip to page 51, near the end of this chapter.

If you will stick with it, I will be your docent on a short excursion through the remarkable world of climate change. You will be able to understand, for example, how in natural climate oscillations, the temperature change must precede the carbon dioxide change. You will also gain a quantitative appreciation of implications for human-made climate change. In return, I hope you will help spread the knowledge. Remember that the fate of our grandchildren depends on a better public appreciation of the situation.

Paleoclimate, especially the waxing and waning of ice ages, is something that you should know about anyhow. Just twenty thousand years ago, most of Canada was under a huge ice sheet, as much as three kilometers (two miles) thick. That ice sheet pushed south, over the U.S. border, covering the areas of Seattle, Minneapolis, and New York.

Ice sheets have continually expanded and retreated for millions of years. While advancing, they shove before them massive amounts of soil—most of the topsoil in Iowa was robbed from Minnesota and Canada by the glaciers. The farmhouse I was born in sat on topsoil so deep that I assumed it went all the way to China. The town I grew up in, Denison, Iowa, is on a hill that is an end moraine, a dirt pile left at the snout of a glacier before it melted.

The size of continental-scale ice sheets is mind-boggling. Although thinner toward the edges, ice over New York towered several times higher than the Empire State building—thick enough to crush everything in today's New York City to smithereens. But not to worry—even though we sometimes hear geoscientists talk as if

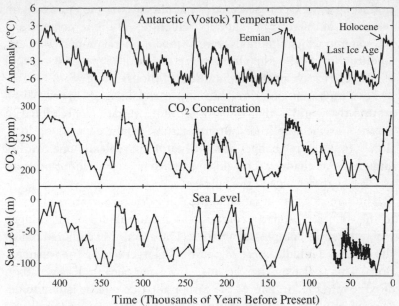

FIGURE 3. *Temperature change, atmospheric carbon dioxide amount, and sea level as a function of time for the past 425,000 years. The horizontal axis shows time in thousands of years before present. Time zero ("present") refers to the date 1750, just before the industrial revolution. (Figure from Hansen et al., "Target Atmospheric CO_2." See sources for chapter 8. For the original data, see sources for chapter 3.)*

ice ages will occur again, it won't happen—unless humans go extinct. Forces instigating ice ages, as we shall see, are so small and slow that a single chlorofluorocarbon factory would be more than sufficient to overcome any natural tendency toward an ice age. Ice sheets will not descend over North America and Europe again as long as we are around to stop them.

Let us look at climate oscillations of the past 425,000 years. The temperature in Antarctica is shown by the top curve in figure 3. Scientists obtained that temperature record by extracting a core (cylinder) of ice from the Antarctic ice sheet, the core extending from the surface all the way to the base of the two-mile-thick ice sheet. The ice sheet was formed by snow that piled up year after year and compressed into ice, and properties of that ice reveal the temperature when the snowflakes formed.

Temperatures at many places around the world are obtained in analogous ways. Ocean sediments that pile up over the years contain the shells of microscopic animals, which reveal the temperature of the water in which the shells were formed. Mineral properties in stalagmites, formed by dripping water in a cave, also preserve a record of temperature changes over hundreds of thousands of years.

The important point revealed by the data from many places around the world is that the large climate variations are global in extent. But the amplitude of temperature change depends on location. Temperature changes at the equator are typically one third as large as polar changes. The global average change is about one half as large as the change at the poles.

The same ice cores that yield the Antarctic temperature allow us to measure atmospheric composition from bubbles of air trapped when the snow compressed into ice. The amount of carbon dioxide, shown in the middle curve in figure 3, is larger during the warm periods. This is as expected, because a warmer ocean releases carbon dioxide into the air. Part of the carbon dioxide release is due to decreased solubility as temperature rises (just as warm soda releases its fizz), and part is due to other mechanisms including reduced storage of biological carbon in the deep ocean as ocean circulation speeds up in interglacial periods.

Close examination shows that temperature changes precede the carbon dioxide changes by several hundred years. Carbon dioxide change in response to climate change is an important feedback process that affects climate sensitivity, as I will discuss momentarily. But note here that the sequence (carbon dioxide change following temperature change) and the delay (several hundred years) are as expected for these natural climate changes. The length of the delay of the carbon dioxide response to temperature change is due to the ocean turnover time, which is several centuries.

When ice sheets melt, the water ends up in the ocean, and sea level rises. The bottom curve in figure 3 shows that sea level changes are large. Twenty thousand years ago, sea level was 110 meters (about 350 feet) lower than it is today, exposing much of the present continental shelves. The rate of sea level rise can be rapid once ice sheets begin to disintegrate. About 14,000 years ago, sea level increased 4 to 5 meters per century for several consecutive centuries—an average rate of 1 meter every 20 or 25 years.

These climate changes surely affected human development. The oldest fossil evidence for anatomically modern humans is from Africa about 130,000 years ago, although the *Homo sapiens* species probably originated about 200,000 years ago. Thus early humans lived during the penultimate interglacial period, the Eemian. As shown in figure 3, the Eemian was slightly warmer than the Holocene, the interglacial period in which we live. Global average temperature in the Eemian was less than 1 degree Celsius warmer than at present, which we will see is an important piece of information in assessing the dangerous level of global warming.

The descent out of Eemian warmth into ice age conditions must have been stressful on humans, even though it took thousands of years. Indeed, the final descent into full ice age conditions 70,000 years ago was rapid and coincided with the one near extinction of humans; as few as one thousand breeding pairs are estimated to have survived during the population bottleneck. A popular theory for the cause of both this rapid cooling and population decline is the colossal eruption of the Toba supervolcano at about that time. Geologic records indicate that Toba ejected at least eight hundred cubic kilometers of material, compared with four cubic kilometers from the 1991 Pinatubo eruption, the largest volcanic eruption of the past century. Regardless of the validity of the Toba theory, it is likely that the rapid global cooling at that time played a role in the population bottleneck.

The huge sea level changes illustrated by the lower curve in figure 3 have played an important role in the development of human societies. Low sea level during the last glacial period produced the Bering land bridge connecting eastern Siberia and Alaska. This grassland steppe region, sometimes called Berengia, was up to a thousand miles wide from north to south. Asians that migrated into Berengia became isolated from ancestor Asian populations. Glaciers that had blocked the path southward began to melt 16,000 or 17,000 years ago, enabling human migration into the Americas.

It was actually the absence of sea level change that helped lead to the development of complex human societies. The social hierarchies of complex societies require food yields sufficient to support the non-food-producing component. Curiously, almost all of the first known population centers, on several continents, date to about 6,000 to 7,000 years ago, when the rate of sea level rise slowed markedly. Until then, as shown in figure 3, sea level had increased continually (not continuously) at an average rate of more than one meter per century for several thousand years. Most human settlements were either coastal or riverine, often in delta regions. Coastal biologic productivity and fish populations are low while sea level is changing, but they can increase an order of magnitude with stable sea level. Thus it has been hypothesized that the high-protein fish diets that become possible with stable sea level account for the near-simultaneous development of complex societies worldwide. This near-simultaneity is surely exaggerated by the fact that earlier settlements were simply flooded or washed away by rising seas. But there is little doubt that our civilizations would have had much greater difficulty getting started, and probably would be less developed today, if sea level had not stabilized. As we shall see, however, the period of near-stable sea level is about to end.

The strong correlation of temperature, carbon dioxide, and sea level is obvious in figure 3. But what are cause and effect? Presumably you would like to know: What causes the huge climate changes? After all, Central Park in summer today is not covered by a kilometer of ice. The culprits, slight perturbations in Earth's orbit around the sun and a tiny tilt of Earth's spin axis, may be surprising if you have not trafficked in this topic. But first I need to clarify the topic of climate sensitivity.

Climate sensitivity was first investigated seriously in 1979, with the help of President Jimmy Carter. President Carter was a worrier—the worrier in chief. One product of his concerns was the *Global 2000 Report*, a several-inches-thick compilation of a huge number of concerns about the future. Considering that Carter initiated and approved projects aimed at extracting oil and gas from coal, as well as cooking the Rocky Mountains to squeeze oil from tar shale, he had very good reason to worry. Those projects, if they had been carried to full fruition and spread to other nations, had the potential to exterminate all life on Earth.

Carter's great contribution to climate science was his request that the National Academy of Sciences prepare a report about the potential climate threat posed by increasing atmospheric carbon dioxide. The academy was established by President Abraham Lincoln in 1863 for just such a purpose: to advise the nation on important matters that required the best scientific expertise. The academy made the perfect choice when it selected Professor Jule Charney of the Massachusetts Institute of Technology to lead this study group.

A lesser scientist might have prepared a report that went into great detail about climate complexities and how climate and carbon dioxide were changing year by year, then made some estimates about how things might continue to change in the future, all with large uncertainty (I am not criticizing the reports of the Intergovernmental Panel on Climate Change—those detailed reports also have a useful place). But Charney chose a very different path. He decided to define a simple, highly idealized problem—a gedankenexperiment—allowing the focus to be on the important physical mechanisms. Thirty years later, Charney's thought experiment has become even more powerful, indeed, an essential element in climate change analysis.

Charney's thought experiment was this: Assume that the amount of carbon dioxide in the air is instantly doubled. How much will global temperature increase? He also specified, at least implicitly, that many properties of Earth should be rigorously fixed, for example, ice sheets and vegetation would remain the same as they are today, and sea level would not change. Only the atmosphere and ocean would be allowed to change in response to the carbon dioxide doubling.

Charney realized, of course, that some of these "fixed" quantities may start to vary on time scales of practical importance. But humans were beginning to burn fossil fuels so rapidly that a doubling of carbon dioxide could be expected in less than a century, which is almost instantaneous on geologic time scales. It was thought that ice sheets would change mainly on millennial time scales. Regardless of the validity of such assumptions, Charney's idealized problem allowed attention to be focused on certain climate processes that are surely important. Just bear in mind that additional processes may come into play over a range of time scales.

Charney was seeking the equilibrium global warming, the warming after the atmosphere and ocean have come to a new final temperature in response to increased carbon dioxide. The immediate effect of doubling carbon dioxide, if *everything* else were fixed, would be a decrease of about 4 watts (per square meter) in the heat radiation from Earth to space. That is simple physics, as explained in chapter 1: The added carbon dioxide increases the opacity (opaqueness) of the atmosphere for heat radiation, so radiation to space arises from a higher level, where it is colder, thus reducing emission to space.

Any physicist worth his salt can immediately tell you the answer to Charney's problem if everything except temperature is fixed. Every object emits heat radiation based on its temperature—if it gets hotter, it emits more radiation. There is a well-known equation in thermodynamics, Planck's law, which defines the amount of radiation as a function of temperature. The average temperature in Earth's atmosphere—about −18 degrees Celsius, or 0 degrees Fahrenheit— causes Earth to emit about 240 watts of heat energy to space, as calculated with Planck's law. If we double the amount of carbon dioxide in the air—as in Charney's thought experiment—that reduces Earth's heat radiation to space by 4 watts, because the carbon dioxide traps that much heat. We can use Planck's law to calculate how much Earth must warm up to radiate 4 more watts and restore the planet's energy balance. The answer we find is 1.2 degrees Celsius. So the climate sensitivity in this simple case of Planck radiation is 0.3 degree Celsius per watt of climate forcing.

This simple Planck's law climate sensitivity, 0.3 degree Celsius for each watt of forcing, is called the no-feedback climate sensitivity. Feedbacks occur in response to variations in temperature and can cause further global temperature change, either magnifying or diminishing the no-feedback, or blackbody, response. Feedbacks are the guts of the climate problem. Forcings drive climate change. Feedbacks determine the magnitude of the climate change.

Curiously, the most important climate feedbacks all involve water, in either its solid, liquid, or gas form. For example, when Earth becomes warmer, ice and snow tend to melt. Ice and snow have high reflectivity, or "albedo" (literally, "whiteness"), reflecting back to space most of the sunlight that hits them. Land and ocean, on the other hand, are dark, absorbing most of the sunlight that strikes them. So if ice and snow melt, Earth absorbs more sunlight, which is a "positive" (amplifying) feedback.

Water vapor causes the largest climate feedback. When air becomes warmer, it can hold more water vapor. Air holds much more water vapor in summer than in winter. Even when snow is falling, which means relative humidity is near 100 percent, if you let the outside air in and warm it to room temperature, you will find that it is exceedingly dry. And air over the Sahara Desert holds a lot of water vapor, even though the relative humidity is low. The reason is that the amount of water vapor air can hold before becoming saturated, thus causing vapor to condense out as water or ice, is a strong function of temperature.

Water vapor therefore causes a positive feedback, because water vapor is a powerful greenhouse gas. Every week or so I get an angry e-mail from somebody seemingly shaking his or her fist, saying something like (with expletives deleted), "What nonsense to say carbon dioxide is important! Water vapor is a much stronger greenhouse gas, and it occurs naturally!" Well, yes, that is so, but the amount of water vapor in the air is determined by temperature. Relative humidity averages about 60 percent. Vapor is continuously provided by evaporation from water bodies, and it is wrung out of the air at times and places where weather fluctuations cause the humidity to reach 100 percent. Thus, when a climate forcing causes global temperature to change, water vapor provides an amplifying feedback.

Is this getting dull, too complicated? Hang on! Soon you will see how the whole feedback problem can be illuminated in one fell swoop. But you need to be aware of the other major feedbacks— since feedbacks determine the magnitude of climate change, and contrarians tie up congressional hearings trying to confuse us about feedbacks.

Here are two examples, briefly. First, clouds. For thirty years the scientific community has been trying to model clouds, trying to understand how the many cloud types will change when climate does. Will there be more clouds, fewer clouds? Will cloud height increase or decrease? We do not even know whether the cloud feedback is amplifying or diminishing.

Second, aerosols (fine particles in the air). Atmospheric dust changes if climate changes. Paleoclimate records (ice cores) show that colder climates are usually dustier. But climate forcing by dust is uncertain, because it depends sensitively on how much sunlight the aerosols absorb and on the altitude of the aerosols in the atmosphere. And dust is just one of many aerosols. Consider dimethyl sulfide, a gas produced by marine algae that forms various aerosols. Algae also change as climate changes, thus changing dimethyl sulfide and its aerosols, thus causing another feedback. Here is a killer: Aerosol changes alter clouds in very complicated ways, because aerosols are condensation nuclei for cloud droplets.

Now you may have an inkling why Vice President Cheney, very politely, asked Ron Stouffer to sit down without finishing his climate-modeling presentation at the first Task Force meeting. Policy makers do not want to try to understand all the feedbacks, especially when we scientists do not yet understand them very well.

It may seem that I am harsh on climate models when I rank their value below paleoclimate studies and ongoing climate observations. But I am not really; I have worked on climate models for more than thirty years. I realize they are needed to help us define which processes are more important, which less so; what observations are needed; and even how we might extrapolate into the future.

Global climate models do a decent job of demonstrating certain feedbacks, such as water vapor and sea ice, even though they failed to predict the recent rapid Arctic sea ice loss. Yet when Jule Charney used existing climate models to estimate climate sensitivity for doubled carbon dioxide, he could say only that it was probably between 1.5 and 4.5 degrees Celsius. And by "probably," he meant that there was only a 65 percent chance that it was in that range.

Thirty years later, models alone still cannot do much better. Here is another killer: Even as our understanding of some feedbacks improves, we don't know what we don't know—there may be other feedbacks. Climate sensitivity will never be defined accurately by models.

Fortunately, Earth's history allows precise evaluation of climate sensitivity without using climate models. This approach is suggested by the fact that some feedback processes occur much more rapidly than others.

For example, water vapor must be a fast feedback, because condensation or evaporation happens quickly after temperature changes. Ice sheets, on the other hand, respond more slowly. It is usually thought that ice sheets require millennia, or at least centuries, to come to a new equilibrium size after a change of global temperature. Thus Charney's idealized problem, with ice sheets, vegetation distribution, and sea level all fixed, can be viewed as an attempt to evaluate the fast-feedback climate sensitivity.

Charney's fast-feedback sensitivity is, by definition, global surface warming after atmosphere and ocean come to equilibrium with doubled carbon dioxide. In reality, some slow feedbacks that Charney fixed by fiat may begin to change before the atmosphere and ocean have come to a new equilibrium. These slow feedbacks, in principle, can be either positive (amplifying) or negative (diminishing). The most startling advances in recent understanding of climate change involve the realization that the dominant slow feedbacks are not only amplifying; they are not nearly as slow as we once believed.

Using Earth's history, we can evaluate Charney's fast-feedback climate sensitivity by comparing the last glacial period, 20,000 years ago, with the recent interglacial period, the late Holocene. We know that, averaged over, say, a millennium, Earth was in energy balance during both periods. We can prove this by considering the contrary: A planetary energy imbalance of 1 watt provides energy to melt enough ice to raise sea level more than a hundred meters in one millennium—but we know that sea level was stable in both periods. The only other place that such energy imbalance could go, other than melting ice, is into the ocean—but ocean temperature was also stable in both periods.

So Earth was in energy balance within a small fraction of 1 watt in both periods. Now we can compare the two periods—two very different climates, both in equilibrium with whatever forcings were acting. Global average temperature was 5 degrees Celsius warmer in the Holocene than in the last ice age, with an uncertainty of 1 degree Celsius.

What factors caused Earth to be warmer in the Holocene? There are three possibilities: (1) a change in the energy received by Earth, that is, a change in the sun's luminosity; (2) changes within the atmosphere; or (3) changes at Earth's surface. We can eliminate the first possibility because while our sun is an ordinary young star, still "burning" hydrogen to make helium by nuclear fusion and slowly getting brighter, in 20,000 years the brightness increase was negligible—0.0001 percent, or about 0.0002 watt. The second and third factors, however, are both important, and they are both accurately known.

We have samples of the atmosphere that existed 20,000 years ago, from bubbles of air trapped in ice sheets. These bubbles reveal that all three of the long-lived greenhouse gases, carbon dioxide, methane, and nitrous oxide, were more abundant during the Holocene than during the ice age. The climate forcing due to these gas changes was 3 watts, with an uncertainty of about 0.5 watt. We also know the changes on Earth's surface from geological data. The biggest change was the large ice sheet covering present-day Canada and parts of the United States and smaller ice areas in Eurasia during the ice age. Changes in vegetation distribution and exposure of continental shelves had smaller effects. The net effect of these surface changes, due to the reduction of the amount of absorbed sunlight during the ice age, was a forcing of about 3.5 watts.

If we add the two together, we see that the total forcing of about 6.5 watts maintained an equilibrium temperature change of about 5 degrees Celsius, implying a climate sensitivity of about 0.75 degree Celsius for each watt of forcing. This corresponds to 3 degrees Celsius for the 4-watt forcing of doubled carbon dioxide. The sensitivity is smack in the middle of the range that Charney estimated, 1.5 to 4.5 degrees Celsius.

The coup de grâce, the slaying of Charney's climate sensitivity beast, is now obtained by considering the entire ice core record. The ice core provides a continuous record of atmospheric composition, while sea level records imply the changing size of continental ice sheets. Thus the climate forcings by atmosphere and surface are readily computed, as shown in the upper part of figure 4. The sum of these two curves, multiplied by 0.75 degree per watt, yields the calculated temperature curve (shown in the lower part of figure 4), which agrees remarkably well with observations.

The most important merit of our empirically derived climate sensitivity is this: All physical mechanisms that exist in the real world are included—and furthermore, they are included correctly; the physics is exact. The resulting uncertainty, or "error bar," on the derived fast-feedback climate sensitivity is small, about 0.5 degree Celsius for doubled carbon dioxide.

I should offer one caveat here. The climate sensitivity we have derived is valid for today's climate and a broad range of climate states. But sensitivity depends on the climate state. Climate sensitivity graphed as a function of mean global temperature forms a U-shape curve. Global temperature today is at the bottom of the U curve, and the bottom is quite flat. But if the planet becomes much colder or much warmer, climate sensitivity will increase; indeed, we will meet the "snowball Earth" and "runaway greenhouse" instabilities. More on those in chapters 10 and 11.

Now we are ready to discuss the drive, the instigation, for the glacial to interglacial climate changes. The climate changes, remember, are enormous—resulting in either a flower garden in Central Park or a kilometer-thick layer of ice. The mechanisms immediately responsible for the entire global temperature change, as we have

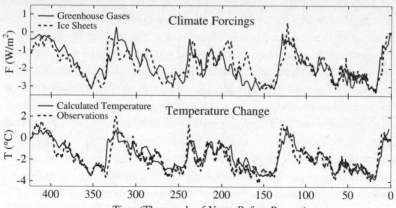

FIGURE 4. *Climate forcings due to greenhouse gas and surface reflectivity changes. Multiplication for the sum of these two forcings by 0.75 degree Celsius per watt yields the calculated temperature. The estimate for observed temperature is Antarctic temperature divided by two. (Figure from Hansen et al., "Target Atmospheric CO_2." See sources for chapter 8. For the original data, see sources for chapter 3.)*

seen, are changes in the amount of greenhouse gas and surface re flectivity. But both these mechanisms are slow feedbacks, not the instigating forcing. The basic mechanisms instigating these changes were suspected for more than a century, and confirmed in the 1970s, but the profound implications are only beginning to be appreciated.

Instigation is provided by small changes to Earth's tilt and orbit around the sun. These changes happen because other planets, especially the heavy ones, Jupiter and Saturn, tug on Earth as they move closer or farther away. The changes have almost no effect on the total amount of sunlight received by Earth averaged over the year. But they do affect the geographical and seasonal distribution of the insolation (a portmanteau of "incident solar radiation"). And they do affect climate in two essential ways.

The simplest effect is due to the change of Earth's tilt, the inclination of the spin axis relative to the plane of the orbit. Today the tilt is about 23.5 degrees and slowly "straightening up." It will reach its minimum tilt, about 22.1 degrees, in about 8,000 years. It takes about 41,000 years to go through the full cycle from minimum tilt, to maximum tilt (24.5 degrees), back to minimum tilt.

The effect of increased tilt is simple: The polar regions of both hemispheres are exposed to greater insolation, while lower latitudes receive less insolation. Increased tilt also causes the amount of summer insolation to increase and winter insolation to decrease. Both the latitudinal and seasonal radiation changes work in the sense of causing high-latitude ice sheets to melt when the tilt increases. Decreased tilt works in the opposite way: Warmer winters yield greater snowfall, and cooler summers increase the chance that snow can survive the warm season, allowing ice sheets to grow. Thus if tilt were the only factor, we would expect Earth to be headed now toward growth of high-latitude ice sheets, possibly toward an ice age, because the spin axis is straightening up.

The second insolation effect is only a bit more complicated. Earth's orbit is slightly elliptical. Earth is now closest to the sun in January and farthest from the sun in July. But the day in the year at which Earth is closest to the sun moves through the entire calendar over the course of approximately 20,000 years (this is caused by Earth's spin axis precessing like a wobbling top, but there is no need to visualize that geometry). Being closest to the sun in January is favorable to building ice sheets in the northern hemisphere, because it makes winter warmer and summer cooler. This second insolation effect works opposite in the two hemispheres, because January is midsummer in the southern hemisphere—so the present situation tends to melt ice in the southern hemisphere.

The second insolation effect is more complicated than the first because it depends on the eccentricity of Earth's orbit—clearly the effect disappears entirely if the orbit becomes circular. The eccentricity of Earth's orbit varies from nearly zero (circular orbit) to almost 6 percent. It is often said that the eccentricity varies with 100,000-year periodicity, but actually it does not have a simple periodicity. At present the eccentricity is quite small, about 1.7 percent, so the second insolation effect is quite weak.

Both insolation effects presently are pushing Earth toward building ice sheets in the northern hemisphere, and thus toward the next natural ice age. Without humans it is not certain whether the present interglacial period, the Holocene, would have ended in the next few thousand years, or whether it would have survived through another precession cycle, similar to the interglacial period 400,000 years ago, which lasted about 40,000 years. The reason to suspect that Earth may have avoided a near-term ice age is that the two in-solation effects are out of phase by about 10,000 years, and the sec-ond effect is very weak because orbital eccentricity is so small.

However, despite the number of scientific papers this matter has generated, the question of when, absent humans, Earth would have headed into its next ice age is about as useful as asking how many angels can dance on a pin. Although both insolation effects now fa-vor ice growth in the northern hemisphere, ice is actually melting rapidly. Human-made climate forcings are now in total dominance over the natural forcings.

The natural forcing due to insolation variations, averaged over the planet, is a small fraction of 1 watt. This very weak forcing is ef-fective only because, operating over long periods, it succeeds in bringing into play two powerful slow feedbacks: global surface re-flectivity changes and greenhouse gas changes. The forcing mecha-nism is seasonal and geographical insolation anomalies, which cause the area of ice and snow in a region or hemisphere to grow or diminish. Although this climate forcing is small, the minor effect on global temperature begins to bring into play the global surface albedo and carbon dioxide feedbacks. Methane and nitrous oxide work in the same way as carbon dioxide, increasing in atmospheric amount as the planet warms and decreasing as it cools. Thus, like carbon dioxide, they are amplifying feedbacks, but smaller ones than carbon dioxide. As shown by figure 4, the global surface albedo and greenhouse gas changes account for practically the entire global climate change.

Both global surface albedo and greenhouse gas amount are now under human control. The slow-feedback processes that cause glacial-to-interglacial oscillations are still operating, of course, but they respond, as they always have, to global temperature. The global cooling trend needed to cause the slow feedbacks that would take Earth into its next ice age no longer exists. Thus any thought that natural processes can still somehow move Earth toward the next ice age is utter nonsense. Humans, by rapidly burning fossil fuels, have caused global warming that overwhelms the natural tendency toward the next ice age. Global temperature always fluc-tuates on short time scales, because of the dynamical sloshings dis-cussed in chapter 1. But human-made climate forcing is now so large that decadal-mean climate will continue to warm for at least the next few decades. Indeed, as we shall see, because of slow feedbacks, global temperature will continue to rise for decades and millennia unless we reduce human-made climate forcings.

The natural climate variations shown in figure 3 have a great deal more to tell us about the future. Note that warmings can proceed quite rapidly, because the disintegration of an ice sheet is a wet process, spurred by positive feedbacks. As the ice sheet begins to melt, it becomes darker, absorbing more sunlight. As the ice sheet's thickness decreases, the surface is at a lower altitude, where it is warmer. There are other feedbacks, both amplifying and diminishing, which we will consider later. Overall, the empirical data show us that natural ice sheet disintegrations can be rapid, at rates up to several meters of sea level rise per century. Sea level fall is usually slower, limited by the snowfall rate in cold places. However, there have been instances—for example, when the meteorology was such that storm tracks consistently drove moist warm air into a region of ice sheet formation—in which sea level fell quite rapidly.

The past seven thousand years of sea level stability is an unusual event. This recent sea level stability occurred because Earth was warm enough to keep ice sheets from forming on North America and Eurasia but cool enough to maintain stable ice sheets on Greenland and Antarctica. The trick that stopped ice sheets from melting seven thousand years ago and kept the sea level almost stable was the slight cooling of Earth from the peak warmth that occurred in the early Holocene. Today, however, global warming of 0.8 degree Celsius in the past century, and of 0.6 degree Celsius in just the past thirty years, has brought global temperature back to at least peak Holocene level, and sea level rise is beginning to accelerate. Sea level is now rising more than three centimeters per decade—double the rate that occurred in the twentieth century.

Accurate measurements of mass being lost by Greenland and Antarctica did not yet exist when I spoke at the White House in 2003, so I used a photograph of surface melt on Greenland to show the kind of processes that could begin to speed climate change (similar to figure 6 on page 78). The photograph shows a small river of meltwater that had formed on the ice sheet and carried water to its base, where it increased the rate of discharge of icebergs to the ocean.

During my presentation, I argued that paleoclimate records provide guidance for the level of warming that would be dangerous from the perspective of sea level change. Specifically, I pointed out that prior warmer interglacial periods such as the Eemian were only about 1 degree Celsius warmer than today, on global average, yet sea level was four to six meters higher than today.

I described the rationale of the alternative scenario paper. If annual fossil fuel emissions level out within this decade and then begin to slowly decline, the CO_2 increase by 2050 would be about 75 ppm. This is a very different scenario than envisioned by government energy departments, which assume that we will keep burning fossil fuels faster and faster, yet it seems conceivable if a constraint on coal is applied, specifically, a requirement of carbon capture and sequestration for future coal plants.

The other half of the alternative scenario involves the non–carbon dioxide forcings. I showed the White House group the bar graph in figure 1 (see page 6) for that discussion. After carbon dioxide, the two largest forcings are methane and black carbon (black soot).

The methane increase since preindustrial time causes about half as much warming as carbon dioxide. The methane warming includes indirect effects of methane on tropospheric ozone and stratospheric water vapor, because these gases are increased via chemical reactions caused by the methane. Methane emissions can be reduced by capture at coal mines, landfills, and agricultural and waste management facilities. The captured methane also has economic value as natural gas. Methane produces carbon dioxide when it is burned, but a methane molecule, with its indirect effects included, is 33 times more potent than a carbon dioxide molecule over a hundred years. Thus the warming effect of methane is reduced 97 percent if the methane is burned instead of released into the air.

Controlling black carbon is more complicated, if the aim is to reduce global warming. Many activities that produce black carbon also produce other aerosols such as sulfates and organic aerosols. These other aerosols are "white," that is, they reflect sunlight and thus have a cooling effect. From a health standpoint, though, all aerosols are presumed to be bad.

My advice regarding black carbon was to place the greatest emphasis on reducing aerosol sources that have a dominance of black carbon over other aerosols. The example that I gave was diesel engines, which are frequently used for trucks, buses, and tractors. The needed regulation was not to ban diesel engines—which are generally more carbon dioxide efficient than gasoline engines—but rather to make emissions standards higher. Engine and "particle trap" technologies could reduce emissions by an order of magnitude with little loss of efficiency.

When Connaughton later testified before Henry Waxman's Committee on Oversight and Government Reform on March 19, 2007, he said that the Bush administration had responded to my recommendations for reducing non–carbon dioxide emissions, including methane and black soot. I do not know that in fact their actions were based on my recommendations. Dina Kruger of the EPA had been advocating methane programs for years, and scientists such as Mark Jacobson of Stanford University had been making a strong case for reducing black soot emissions. However, Kruger mentioned to me in 2003 that the interest of the White House in these non–carbon dioxide climate forcings permitted significant programs to move forward.

One resulting action was a methane-to-markets program that helps reduce methane emissions via capture at coal mines, landfills, and agricultural and waste management facilities and uses the captured methane as fuel. White House interest helped Kruger and the EPA initiate the program in the United States and extend its effectiveness via cooperation with several developing countries that have larger methane emissions than the United States. This approach, extended globally, is better than the Kyoto Protocol approach, in my opinion. Methane is one of the escape hatches that make the Kyoto approach ineffectual for carbon dioxide. The Bush administration also deserves credit for major tightening of soot emission limits in the face of opposition from diesel producers, truckers, and other industries. In addition to supporting rules that reduced soot emissions from trucks and buses, the administration later expanded regulations to cover tractors, trains, and ships.

So, it seems to me that Connaughton is justified in his assertion that he, CEQ, and the administration have been responsive to recommendations regarding the non–carbon dioxide climate forcings. With regard to the more important issue, carbon dioxide, it is not clear how much possibility there was for Connaughton to influence the policies set by the president and vice president. He was in a position to make recommendations. But did I or anybody else make a case strong enough for him to be able to challenge decisions of Cheney and the Task Force?

During my presentation, I showed graphs for fossil-fuel-use scenarios designed to keep additional global warming less than 1 degree Celsius (the alternative scenario) and less than 2 degrees. I argued, based primarily on paleoclimate sea levels, that we should aim to keep warming less than 1 degree. But I had to admit that some other scientists, with more of a background in ice sheet physics and paleoclimate studies, were arguing that a 2-degree-Celsius limit was appropriate. Also, it must have been apparent that my arguments about the potential instability of ice sheets during "wet" disintegration involved a good deal of "arm waving"—a qualitative argument without much quantitative backing.

The friendly discussion with Connaughton and the other participants from CEQ, OSTP, and OMB left me with positive impressions of the group and individuals that I spoke with. Connaughton, in particular, was interested in a number of things discussed—he impressed me as being smart (his biography notes that he received a B.A. from Yale University and graduated second in his class, magna cum laude, Order of the Coif, from the Northwestern University School of Law). Often in talking with people in high positions I can sense that they have rigid positions they are defensive about, but I did not get that impression about Connaughton. Anniek tells me that I always wear my opinions and feelings plainly on my face, and warns me that most people do not do that. She is right that I am not good at reading people, but I saw no hint that Connaughton was Machiavellian.

I was escorted from the briefing by David Halpern, a staff member in Marburger's OSTP, who commented that they would have another presentation the following week on the same topic—by Richard Lindzen. Hmm, I thought, as any faint hopes of a change in policy quickly began to dim. I declined an invitation to return to Washington for Lindzen's talk because I had other obligations, but I later received a copy of Lindzen's presentation from Halpern. I left a copy of my presentation with OSTP, and I also made the presentation available on the Internet.

Lindzen's presentation warrants comment, because U.S. policies regarding carbon dioxide during the Bush-Cheney administration seem to have been based on, or at minimum, congruent with, Lindzen's perspective.

My hope was that Lindzen would address fundamental scientific issues. Climate sensitivity—long-term global warming in response to a specified climate forcing—had long been the main issue. Charney recognized that climate sensitivity must be the first question addressed. For years Lindzen had insisted that climate sensitivity to doubled carbon dioxide could be no more than a few tenths of a degree. So here was a clear disagreement by an order of magnitude, that is, by about a factor of 10. Best of all, Lindzen's position was documented in publications including his own testimony to Congress and in a few papers published in scientific journals, as well as in a summary published by Richard Kerr in the December 1, 1989, issue of *Science* magazine.

I realized that climate sensitivity was in the process of being nailed down—rigorously and accurately defined by the paleoclimate information discussed in this chapter. Of course, even today it is possible to find scientists and published papers concluding that climate sensitivity is quite uncertain. A common approach is to calculate the expected warming of the past century based on assumed climate forcings—then, because of uncertainties in actual forcings, conclude that climate sensitivity is only constrained to lie somewhere within a large range, say 2 to 8 degrees Celsius for doubled carbon dioxide. That logic is a case of failing to see the forest for the trees. Our knowledge is not based on the dullest instrument in our tool bag. Rather it is based on the sharpest, most discriminating information we can muster.

Obtuseness concerning climate sensitivity reminds me of a story Richard Feynman told about his early experience at Los Alamos, where many of the top physicists in the world had assembled to work on the Manhattan Project. Feynman would eventually become known, at least among many physicists, as the second greatest scientist of the twentieth century, but at that time he was just becoming a postdoc. He finished his Ph.D. thesis in a rush, as he and many of the top physicists essentially dropped everything to join the effort to help build an atomic bomb (after Einstein warned President Roosevelt, in a now-famous letter, that Germany was probably working on a bomb of monstrous power).

Feynman was at a meeting of some of the physics giants—Richard C. Tolman, Arthur Compton, Isidor Isaac Rabi, J. Robert Oppenheimer, and others—and they were talking about the theory of how they were going to separate uranium. Feynman understood the specific matter being discussed pretty well. Compton explained one point of view, which Feynman could see was right. But they went around the table, someone saying there's a different possibility, another suggesting still a different idea, all the time Feynman becoming more and more antsy and jumpy, thinking, why didn't Compton repeat his argument? After they had gone around the table, Tolman, the chairman, said something to the effect of "Well, it's clear that Compton's argument is best, and now it's time to go ahead." It made a big impression on young Feynman—seeing how really good scientists work. They wanted to look at a problem from all angles, reexamining alternatives and different facets, to guard against a mistake. All the while they could recognize the best idea without having to repeat the arguments.

Jule Charney, were he alive today, would be thrilled by the paleoclimate information on climate sensitivity. Undoubtedly he would stand up and say, "Great, now let's move ahead." Dick Lindzen is a whole different kettle of fish. He has made numerous scientific contributions, received significant honors, and suggested interesting ideas. But as for an overview and insight about how climate works, he is no Jule Charney by any means. Lindzen's perspective on climate sensitivity, as he told Richard Kerr, stems from an idea of a theological or philosophical perspective that he doggedly adheres to. Lindzen is convinced that nature will find ways to cool itself, that negative feedbacks will diminish the effect of climate forcings. This notion spurred Lindzen to propose a specific mechanism for how the atmosphere takes care of itself: He suggests that columns of tropical cumulus convection intensify if carbon dioxide increases, piping energy high into the atmosphere, where the heat would be radiated to space. This mechanism, he suggests, is nature's thermostat, which keeps global warming at a few tenths of a degree for doubled carbon dioxide, rather than a few degrees.

Charney would understand very well that if the real world possesses such a negative feedback, its effect is included in the empirical sensitivity extracted from Earth's paleoclimate history. A reliable, accurate evaluation of climate sensitivity now exists, including all feedbacks. Is Lindzen likely to admit that he's wrong? Probably not. I expect him to keep asserting that human-made climate change is unimportant on his deathbed, defending that position as a lawyer defends a client. A lawyer does not seek truth; a lawyer seeks a win for a client. That approach makes it difficult for the public. Lindzen makes qualitative statements that sound reasonable, and he raises technical matters that a layperson cannot assess, making it sound like there is an argument among theorists.

Abraham Lincoln, as I have noted, established the National Academy of Sciences for the purpose of providing advice on technical matters. President Bush, early in his first term, asked the academy for advice on global warming. Specifically, the White House sought the academy's evaluation of the conclusions reached by the Intergovernmental Panel on Climate Change. IPCC, in its most recent report, made increasingly strong statements about the likely consequences of continued increases of greenhouse gases. The White House was probably hoping that the academy would document some criticisms of the IPCC report. If so, the White House was disappointed. The academy's evaluation had only mild reservations, giving the conclusions of IPCC strong endorsement overall.

Lindzen, though, had an explanation for the academy report. His talk at the White House was titled "Getting Serious About Global Warming." His first chart, titled "The UC Irvine Atmospheric Chemist Gambit," claimed that Ralph Cicerone and Sherwood Rowland had inserted an "irrelevant opener" into the executive summary of the academy report after the text of the report itself had been agreed to by all participants. Cicerone was chairman of the committee, Rowland was one of the principal authors, and Lindzen and I were two of the other nine committee members.

Cicerone is one of the most respected scientists in the United States and is now president of the National Academy of Sciences. Rowland won the Nobel Prize in chemistry, with Mario Molina and Paul Crutzen, for his prediction that chlorofluorocarbons could destroy stratospheric ozone. Rowland's prediction was validated by nature with the appearance of the Antarctic ozone hole. Because of the warning issued by these scientists, the heavily populated northern hemisphere was largely spared the consequences of ozone depletion.

The paragraph that Lindzen objected to, the first paragraph of the executive summary, reads:

> Greenhouse gases are accumulating in Earth's atmosphere as a result of human activities, causing surface air temperatures and subsurface ocean temperatures to rise. Temperatures are, in fact, rising. The changes observed over the last several decades are likely mostly due to human activities, but we cannot rule out that some significant part of these changes is also a reflection of natural variability. Human-induced warming and associated sea level rises are expected to continue through the 21st century. Secondary effects are suggested by computer model simulations and basic physical reasoning. These include increases in rainfall rates and increased susceptibility of semi-arid regions to drought. The impacts of these changes will be critically dependent on the magnitude of the warming and the rate with which it occurs.

Lindzen's charge—that changes were made late in the reviewing process and that the changes affected the essence of the report's conclusions—sounds serious to the public. But all other committee members know it is nonsense. The report was done quickly at the request of the White House, but editing was done openly, with continual e-mail exchanges with committee members. The public has access to both the executive summary and the full report and can verify that the summary reflects the report's contents—but how many will do so? Instead, the public hears a "balanced" perspective: Many scientists agree that humans are altering climate, but there seems to be disagreement about that conclusion within the scientific community.

Lindzen's perspective is summed up in the final statements on his concluding chart shown at the White House meeting, which read:

> (1) Scientists who are willing to speak out in support of hysteria are supported with funding, awards, and even legal assistance. (2) The environmental movement coordinates public pronouncements so as to guarantee that all spokesmen are "on the same page." (3) Institutions, dependent on

support, are supportive of alarmism. (4) Scientists who protest alarmism are out in the cold. There is no assistance from any direction.

(These statements are similar to his conclusions at the second meeting of Vice President Cheney's Task Force.)

Between his opening chart attacking the integrity of Cicerone and Rowland and his final chart's conclusions, Lindzen's presentation consisted of a criticism of IPCC and "alarmism." As an antidote to such "nonsense," he recommended a book, *Taken by Storm*, by Christopher Essex and Ross McKitrick. Lindzen's presentation included only two scientific graphs. The first graph, from the 2001 IPCC report, showed that climate models using a combination of natural and anthropogenic forcings did a reasonably good job of reproducing global warming of the past century. He criticized this result, because, he said, we don't really know the human-made aerosols and we also don't know the El Niño and volcano forcings. He is right that the human-made aerosols are not measured. But El Niño is not a forcing—as explained in chapter 1, it is an unforced climate variability, a "sloshing" that is unpredictable except on short time scales, and it has little effect on century time-scale climate change.

Lindzen's second graph showed that there was a high correlation between sunspots and the number of Republicans in the Senate. He concluded that the IPCC analysis was hardly better than the sunspot-Republican analysis, indeed, "in some respects the climate analysis is more questionable, since the effect is so much smaller."

Any levity from Lindzen's presentation dissipates upon the realization that his presentations were taken seriously by the administration. There are reasons to believe that Bush, Cheney, and Rove all shared Lindzen's perspective (consistent with evidence presented in chapter 7) and distrusted the scientific community. The answer that the National Academy of Sciences had delivered in response to the president's request, the report that Lindzen "critiqued," was not the answer the White House wanted to hear. The president did not ask the academy for advice about global warming again during the remainder of his eight years in power.

CHAPTER 4

Time Warp

I SAID THAT MY STORY WOULD cover only the past eight years. Sorry. In this chapter I'm going to have to take you with me for a moment into a backward time warp. If you are irascible by nature, easily angered by broken promises, you may wish to skip directly to the next chapter. But in so doing, you will miss a discussion of some potentially crucial information, key to understanding the task of restoring Earth's energy balance—and restoring Earth's energy balance is the fundamental requirement for stabilizing our climate.

It is Kathie Olsen, associate director of the Office of Science and Technology Policy, who is responsible for pulling us into this backward time warp, with a question she asked at the end of my presentation to the White House Council on Environmental Quality in June 2003. Her question was about atmospheric aerosols, the fine particles in the air: What aerosol measurements were needed to define the climate forcing by aerosols and why were the measurements not being obtained?

It was a good question. Another Richard Feynman story can help me explain the answer. The story involves giants in the world of physics, but it has relevance to us ordinary people. You can call it "When Speaking to Authority."

The great Danish physicist Niels Bohr and his son Aage visited Los Alamos at the time when everyone there was working on the bomb. Niels Bohr had won a Nobel Prize in 1922 for his work on the structure of atoms (Aage won his own Nobel Prize, in 1975, for work on the structure of the nucleus). While they were visiting the secret Los Alamos project, Niels and Aage were given the aliases

Nicholas Baker and Jim Baker, but everyone knew who they were—Niels Bohr was a god even to the other famous physicists. At a meeting held to talk about problems with the bomb, Feynman had to sit way in the back because everybody else had crowded close to the great Niels Bohr. The day before the Bohrs were to return for a second visit, Feynman got a call. "Hello, Feynman? This is Jim Baker, my father and I would like to meet with you."

A surprised Feynman said, "Who, me? I'm just . . ." They met at eight A.M., while the other scientists were still in bed. The Bohrs had ideas about how to improve the bomb: "We have this idea, blah, blah, blah." Feynman responded, "No, that won't work because blah blah." "Well, how about blah blah." "That may be better, but it still has this damn fool idea in it . . ." The discussion went on for about two hours, until Niels Bohr lit his pipe and said, "I guess we can call in the big shots now." Aage then explained to Feynman that after their first visit, his father had told him, "Remember that little fellow in the back? He's the only one who's not afraid of me and will say when I've got a crazy idea. So next time that we want to discuss ideas, we'll talk with him first."

Reticence exists in different forms. The reticence I'm concerned with in this chapter is the reluctance to contradict authority. A good scientist is interested in how things work and doesn't want to worry about authority. Niels Bohr might have appreciated respect—but not to a degree that would inhibit discussion. Reticence does not fit the scientific method.

How the real world works is an almost infinitely complex puzzle. A scientist's task is to try to figure out a valid description of some part of the puzzle. If he keeps two sets of books, one he believes and another to please authorities, it makes the problem much harder. So a scientist should be clear and blunt about what he thinks, even if the authorities don't like it—otherwise he will not do very well in science.

Feynman spoke out without reticence. It worked for him. But Feynman had advantages we ordinary people do not have: He was a genius and a scintillating, entertaining communicator. Also, as Feynman frequently acknowledged, he was fortunate to work in places, mostly at universities, where speaking openly was expected and appreciated.

I have had more difficulty. I try to speak up if something seems important. But I have always been shy, a poor communicator, and lacking in tact. I decided that the best chance for me to communicate better was to learn to write better. So for years, after Anniek and I would go to bed, I would read out loud to her, usually English novels, marking words to study later. It improved my vocabulary, but not my tact.

"Get to the point!" you might be thinking. "What is the relevance of communications and tact?" Well, the answer to Kathie Olsen's question about why we do not have the data we need—about climate forcing by aerosols, which requires measuring aerosols and their effect on clouds—turns out to be a story of failed communications.

One way for me to tell that story would be to give a detailed accounting of my efforts over several years, as a scientist, to promote the aerosol and cloud observations that I believe to be necessary. It is a sorrowful tale, but one that will need to wait until I have retired.

Instead, we can look at the science of planetary observations, which was clear before I became a government employee in 1972 and remains valid today.

If you want to make measurements of a planet to learn what its atmosphere is made of and, if the atmosphere is not too thick, learn something about its surface, how can you go about this? Well, unless you plan to fly to the planet, land on it, and start poking around, about the only thing you can do is measure the radiation coming from the planet. Then you have two choices: measure the sunlight reflected by the planet or measure the heat radiation emitted by the planet. You had better choose to measure both if you want to figure out much about the planet.

The reflected sunlight is what you see with your eyes when you look at the planet. The heat radiation needs to be measured with an instrument. In fact, you had better use instruments to measure both the reflected sunlight and the planetary heat radiation if you want to obtain detailed information about the planet.

Figure 5 shows what that radiation "looks like" to a physicist. This is a graph of the radiation intensity (or brightness) as a function of the wavelength of the radiation for a planet of special interest: Earth.

The left graph has two jagged curves. The top curve is the sunlight (solar radiation) that hits Earth, and the lower curve is the amount of sunlight that gets to the ground on a cloudless day. Not all the sunlight reaching the planet gets to the ground—some is absorbed by gases such as water vapor and ozone, and some is reflected back to space by aerosols and air molecules.

The jagged curve in the right graph represents the heat (thermal radiation, or terrestrial radiation) emitted by Earth. This measurement was made by an instrument called IRIS (Infrared Interferometer Spectrometer), which was developed by Rudy Hanel, a NASA scientist, for planetary studies. The measurements shown in figure 5 were made in 1970 above the Sahara Desert, when the instrument was on a satellite orbiting Earth.

Sunlight and Earth's heat radiation bear many similarities. They are both thermal (heat) radiation. It's just that sunlight is coming from a much hotter body. The temperature of the sun's surface is almost 6,000 degrees Kelvin (about 5,700 degrees Celsius). The average temperature of Earth's surface is about 288 degrees Kelvin (15 degrees Celsius), which means it is cooler than the sun by a factor of about 20.

An aside: A Kelvin degree is the same as a Celsius degree, except that zero degrees Kelvin is defined as the temperature at which all molecular motion ceases. That is really "stone cold." Zero degrees Kelvin is about –273 degrees Celsius. That is the coldest anything can be, so zero degrees Kelvin is absolute zero, and temperature in Kelvin is also called absolute temperature. The Kelvin scale might have been chosen "at the beginning" if this had all been understood then. But it is useful to also keep Celsius, which is almost the same as

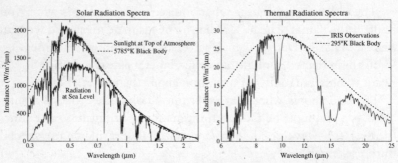

FIGURE 5. *Sunlight reaching Earth and the amount reaching the ground under typical clear-sky conditions (left) and thermal (heat) radiation emitted by Earth (right). (Data from Hansen et al.,* Long-Term Monitoring of Global Climate Forcings and Feedbacks. *See sources.)*

Kelvin except for a constant offset, because who would want to say, "The temperature today will be 288 degrees"; better to have small, more manageable numbers. (The Fahrenheit temperature scale used in the United States, like inches, feet, yards, and miles for distance, instead of centimeters, meters, and kilometers, perhaps should be abandoned, but such decisions are made in Washington. Maybe that's the way it should be in a democracy. But our grandchildren are being handicapped a bit—science would be easier for them if they could use the more logical scales that are used in most other countries.)

The amount of radiation emitted by a body is, approximately, a simple function of its temperature. We have a name for the equation that describes the amount of radiation as a function of temperature and wavelength: the Planck function. It accurately describes the radiation from a perfect absorber, or blackbody, material that completely absorbs all wavelengths of incident radiation, that is, all the energy shining upon it. Black carbon (black soot) is a good approximation of a perfect absorber.

The dashed lines in figure 5 represent the amount of radiation that would be emitted by blackbodies at the temperature of the sun's surface (left graph) and Earth's surface (right graph). The jagged curves for actual radiation measured differ from the ideal blackbody curves because of the absorption of the radiation by gases in either the sun's upper atmosphere or Earth's atmosphere.

Gas absorption occurs at specific wavelengths that depend on the type of gas—so the absorption lines serve as spectral fingerprints that identify gas species. For example, the broad feature at wavelength 15 microns (a micron is a micrometer, one millionth of a meter) in Earth's thermal emission is due to absorption by carbon dioxide. The narrower absorption near 10 microns is absorption by ozone.

Energy absorbed by these gases is promptly reemitted in all directions, but the amount and spectral (wavelength) distribution of the emitted radiation depends on the temperature at the location of the absorbing gas molecules. Because Earth's temperature gets colder the higher we go in the lower atmosphere, absorption by the greenhouse gases reduces the amount of heat radiation to space. Therefore, if the amount of these gases increases, terrestrial radiation to space is reduced. This change causes a temporary planetary energy imbalance, with Earth emitting less energy to space than it absorbs from the sun. So Earth warms up until energy balance is restored.

Thus figure 5 gives a realistic "picture" of the greenhouse effect, which we've already discussed in words.

Our interest here is the information about Earth's atmosphere that we can extract from the planet's thermal spectrum and reflected sunlight. Clearly, the thermal spectrum tells us what gases are in the planet's atmosphere, because each gas has its own spectral absorption signature, as we noted for carbon dioxide and ozone. In addition, we can use this spectrum to measure the vertical temperature profile—how the temperature varies at different altitudes in the atmosphere. Careful measurement of the thermal spectrum reveals many narrow absorption lines. The depth of an absorption line depends on the temperature profile in the atmosphere. If we measure the depth of many lines, very accurately, and compare them, we can deduce the atmosphere's temperature profile.

The amount and accuracy of information that can be extracted from the thermal spectrum depend primarily on the precision with which the radiation intensity is measured at one wavelength relative to the intensity at other wavelengths. It is for this reason that an interferometer measurement, such as those taken by the IRIS instrument and shown in figure 5, is required for the most precise results. In such an instrument a wide range of wavelengths is recorded on the same detector, which allows greater precision than is possible with an instrument that uses separate detectors to record different wavelengths, because the detectors must be calibrated against each other—and calibrations are always imperfect.

Now let's consider the sunlight, shown in the left graph in figure 5. Thermal radiation emitted by either the sun or Earth is practically unpolarized. What does that mean? Well, let's go back a step—all the way back to Isaac Newton in his 1704 book *Opticks*. Newton wrote, "Do not all fix'd Bodies, when heated beyond a certain degree, emit Light and shine; and is not this Emission perform'd by the vibrating motion of its parts?"

Newton had the picture basically right, even though he lived long before there was knowledge of the structure of atoms or the nature of electromagnetic radiation. Oscillating charged particles in any molecule eventually emit a packet of electromagnetic radiation—so-called because it radiates from electrically charged particles. The radiation has the form of self-propagating waves with electric and magnetic components that oscillate perpendicular to the direction of energy propagation. True for sunlight. True for any thermal emission.

If you look at the sun through a polarized lens, which transmits light vibrating in one preferred direction, and rotate the lens, the intensity of radiation does not change. The light is unpolarized because the thermal emission, from the sun in this case, includes many packets of radiation vibrating in random directions in the plane perpendicular to the propagation direction.

However, when sunlight is reflected from a surface or scattered by a particle, it can become polarized. That is the reason for wearing polarizing sunglasses—reflection from a water or road surface is polarized. The water or road reflects mostly the radiation packets vibrating horizontally. So the polarized lenses in sunglasses are oriented vertically to cut down the glare.

The polarization of light scattered by small particles—aerosols or cloud drops—contains an enormous amount of information on the nature of the scattering particles, if the polarization is measured to high accuracy. In 1969 I studied as a National Science Foundation fellow at the University of Leiden Observatory in the Netherlands, under the world's leading expert on light scattering, Professor Henk van de Hulst, and worked also with his top protégé, Joop Hovenier. We showed that the polarization reveals the aerosol amount, the size and shape of the aerosols, and even their index of refraction. This latter quantity—a measure of the angle at which a light ray is bent when it enters the particle—helps to identify aerosol composition.

A basic understanding of the information about reflected solar and emitted thermal radiation illustrated in figure 5 tells us what we need in order to determine the aerosol climate forcing. The required observations are (1) polarization of reflected sunlight to an accuracy of about a tenth of a percent, with a given spot on the ground looked at from several different directions as the satellite passes overhead, and with the measurements made at several wavelengths spread over the solar spectrum; and (2) infrared emission measured with a high-precision interferometer, that is, with an instrument that gives the best wavelength-to-wavelength precision.

Okay, so in 1970 the physics was already clear. The information content on aerosols and their effect on clouds required polarimetry of reflected sunlight and interferometry of thermal emission. But measurements were not started then—climate change was not an urgent issue in 1970. My scientific interest at that time was with other planets.

Almost twenty years later, climate was an issue. In December 1989 I received a letter from senators Al Gore and Barbara Mikulski inviting me to a "roundtable" meeting in Gore's office. They wanted to discuss three proposed programs, all seeking support from U.S. taxpayers: the NASA Earth Observing System, the U.S. Global Change Research Program, and the International Geosphere-Biosphere Program.

I was asked to participate as a scientist, not as a representative of NASA. None of the other participants (James Baker, Francis Bretherton, Tom Lovejoy, Gordon MacDonald, Mike McElroy, Irving Mintzer, Bill Moomaw, and George Woodwell) were government employees. It was not unusual for me to be the only government employee in science advisory meetings—I long had a reputation for giving frank scientific opinions, without concern for institutional implications.

The day of the meeting happened to be the coldest day in the eastern United States in several years. The heating system in the Russell Senate Office Building faltered that day, and it was freezing in the meeting room. As I walked in, Al Gore said, "Say, aren't you the guy who . . ." cutting off his sentence at that point. Everybody had a good laugh. Gore put on his jacket and instructed his staff to bring in pots of hot coffee.

Al Gore was remarkable. He asked questions around the table about the major scientific issues in earth sciences. He was also the note-taker. Every now and then he would say, "Okay, here is what I understand," repeating the essence in language that an educated person could understand. He did a better job than most scientists could do—certainly better than I could do.

A second meeting occurred in Senator Gore's office in January 1990. The invitation letter included a list of twenty-three scientific topics that came out of the first meeting. After seeing that long list, I was determined to raise a fundamental matter at the second meeting: the need for a scientifically defined focus for an observing program.

I already had criticized the planned Earth Observing System at the first public meeting about the project earlier that year; my criticisms were reported in *Science* magazine on June 16, 1989. My concern was in part with the proposed giant observational platforms, dubbed by some as Battlestar Galacticas. Each would cost more than a billion dollars. There was a danger that their size would squeeze out science; their cost would take money better spent on students and postdocs, exclude small satellites for specific long-term measurements, and eliminate the possibility of a quick observing response to new information—the way good science usually works. In the attempt to include a bit of something for everyone on a large platform, basic scientific requirements could go unmet. Indeed, I could see that instruments capable of making measurements needed to understand climate change were not adequately included.

There is a valid scientific rationale for putting large instruments in space to observe climate and other processes on Earth in fine detail. But, in addition, there are highly precise measurements needed to understand long-term climate change that must be continued for decades. For example, human-made sources of aerosols change slowly over the years. The aerosol changes need to be measured, as well as the effect of aerosol changes on clouds. But clouds change for other reasons too—perhaps in concert with solar irradiance changes. So measurements must be continued over at least a couple of ten-to-twelve-year solar cycles.

I went to the second Gore-Mikulski meeting with a table summarizing the measurements needed to analyze long-term climate change (see the table in appendix 2 on page 281). Climate forcings that must be precisely monitored are solar irradiance, greenhouse gases, aerosols, and surface properties. Climate feedbacks are clouds, water vapor, and surface ice and snow. Some of the quantities are both forcing and feedback, but with precise global measurements over sufficient time, it is possible to sort that out.

One conclusion was that all essential measurements could be made by four instruments, each moderate in size and cost. Two of the instruments, the polarimeter, measuring reflected sunlight, and interferometer, measuring thermal emission, would need to be on the same satellite, doing their measurements more or less simultaneously while looking at the same location. A third instrument would precisely monitor the sun's irradiance. A fourth instrument would make precise measurements of aerosols and gases in tenuous higher layers of Earth's atmosphere by observing the sun from a satellite through Earth's atmosphere at sunrise and sunset.

I did not get far with this topic at the Gore-Mikulski meeting. After several sentences, Senator Gore politely interrupted, saying, "With all due respect, Dr. Hansen . . ." It was the problem that Ron Stouffer had had at Vice President Cheney's Task Force meeting—as the emperor in *Amadeus* says to Mozart: "Too many notes." Gore was right: I did not have a clear, succinct story. Besides, he had asked us to focus on any of the twenty-three problems that were amenable to progress within a few years. In Washington, short time scales are emphasized, and I could not promise anything that fast.

I knew that getting support for taking these measurements would not be easy. Before the second Gore-Mikulski roundtable meeting, Michael McElroy, chairman of earth and planetary sciences at Harvard University, pulled me aside. McElroy related that he had recently met the head of the Earth Observing System program, Shelby Tilford, who seemed to be aware of the criticisms of that program expressed at the first Gore-Mikulski meeting. McElroy asked him, "Did you have a mole at the meeting?" and received a flushed response: "Watch it, McElroy, or you will end up in the same box as Hansen."

The anger at me was probably because of my public comments that had been reported in *Science* magazine, rather than what I had said at the Gore-Mikulski meeting. But it seems my untactful communication had sparked the strong reaction—a response I had not expected, given my background in planetary and space science. The space science community did not seem to object to criticisms, but rather used them as a basis for discussion, and scientists had more control of what observations were to be made. In earth science, on the other hand, satellite observations seemed to be organized with a more top-down management, and criticisms were not always appreciated.

It seemed to me that the communication problem could be overcome. Almost every scientist told me privately that I was right about the need to obtain this precise long-term data from small satellites. So for several years I continued advocating this concept to measure the key climate forcings and feedbacks. In 1992, with colleagues Bill Rossow and Inez Fung, I published a comprehensive workshop report that described the science rationale. The workshop included the participation and support of a large number of the best relevant scientists in the country.

Yet despite this documentation, the widespread agreement about the validity of the rationale, and the occurrence of several reassessments of the U.S. climate observing system, I never succeeded in getting the measurements started. The principal reasons given were these: (1) *Do not let the perfect be the enemy of the good*—measurements will be obtained even if they are not as accurate as desired; (2) *the train has left the station*—it would be counterproductive to take a different approach; and (3) *we will improve the system later.*

Although those were the explanations I was often told, my opinion, based on many years of observing the process, is that there was another, more powerful reason behind the scenes. I refer to the special interests—the individuals, organizations, and industries that obtained support from the program focused on large satellite systems. Whatever the reasons, the result is that, twenty years later, we still do not know the aerosol climate forcing or how it is changing.

These same arguments come up as a rationale for ignoring policies proposed to alleviate climate change (as I will show in later chapters). This is happening in discussions both within the United States and in the United Nations, where the objective is to find an effective treaty to succeed the Kyoto Protocol.

In both the United States and the United Nations, the real forces at work have little to do with perfect versus good or trains having already left the station. Indeed, those arguments are readily shown to be bogus. But the story is not a simple one with easily identified good guys and bad guys. For example, someone with the noblest of objectives may feel that he has found a way to work the post-Kyoto system so as to benefit his specific noble objective. In that case, he may be willing to support a system that has no chance of stabilizing climate, perhaps thinking we can improve the system later.

I do not care much whether you try to understand polarimeters or interferometers. But I care a lot whether you understand policy discussions that are going on in Washington and other capitals around the world. If we let special interests rule, my grandchildren and yours will pay the price.

CHAPTER 5

Dangerous Reticence: A Slippery Slope

HUMANITY TREADS TODAY ON A SLIPPERY slope. As we continue to pump greenhouse gases into the air, we move onto a steeper, even more slippery incline. We seem oblivious to the danger—unaware how close we may be to a situation in which a catastrophic slip becomes practically unavoidable, a slip where we suddenly lose all control and are pulled into a torrential stream that hurls us over a precipice to our demise.

You may say, "Surely you are joking, Mr. Hansen!" Would that I were. Human-made climate change is, indeed, the greatest threat civilization faces. Skepticism at such an extreme statement is understandable. The number of degrees Celsius involved in global warming seems small compared with day-to-day temperature fluctuations. How can the warming of the past century, about 0.8 degree Celsius (about 1.5 degrees Fahrenheit), be so important? Even if the warming increases to several degrees this century, as will occur if we continue business-as-usual increases in fossil fuel use, how can warming of several degrees destroy civilization?

The paradox of global warming, the fact that mild heating can have dramatic consequences, first occurred to me one summer day in 1976 as Anniek and I were driving home with our son, Erik, after spending the afternoon at Jones Beach on Long Island. When we had arrived at the beach near midday, we needed to find a spot near the water to avoid the scorching hot sand. Yet by late afternoon it became very cool as a strong wind from the ocean whipped up whitecaps. Erik and I had goose bumps as we ran along the foamy shoreline and watched the churning waves.

Earlier that year Andy Lacis and I, and three colleagues, had calculated the climate forcing by all human-made greenhouse gases. Their heating of Earth's surface had reached a level of almost 2 watts per square meter. The paradox to me as we ran along the beach was the contrast between nature's awesome forces and this seemingly feeble heating. It was hard for me to see how the warmth of two tiny 1-watt bulbs over each square meter could command the wind and waves or smooth our goose bumps. And wouldn't any such low-wattage heating of the ocean surface be quickly dissipated to great depths?

Ah, but remember, climate change between the present interglacial period and the ice age 20,000 years ago was maintained by a forcing of only about 6.5 watts—yet that forcing produced a different world, with Canada and parts of the United States under a thick ice sheet, and a sea level 350 feet lower than it is today. Moreover, the forcings composing the 6.5 watts—3.5 watts from a change of surface reflectivity and 3 watts from a change of atmospheric gases—were, in fact, slow climate feedbacks instigated by a far weaker forcing that was much less than 1 watt on global annual average: the perturbations of insolation on Earth's surface due to small changes in Earth's orbit.

"Well, okay," you may say, "but if it takes twenty thousand years for big changes to occur, why should I care? My children, grandchildren, and I will be long dead, and who knows what else will happen in the interim." The reason is this: Climate change in response to human-made forcing will be much more rapid than these natural changes. The speed of glacial-interglacial change is dictated by 20,000-, 40,000-, and 100,000-year time scales for changes of Earth's orbit—but this does not mean that the climate system is inherently *that* lethargic.

On the contrary. Human-made climate forcing, by paleoclimate standards, is large and changes in decades, not tens of thousands of years. It is important to determine the response time to this forcing via analysis and understanding of the climate system. Unfortunately, paleoclimate provides no known empirical data on the response time for a large, rapid, positive (warming) forcing that perseveres. Volcanic eruptions and asteroid impacts cause rapid and large climate forcings, but those forcings are negative (cooling) and brief. The saw-toothed climate response (shown in figure 3 on page 37) to symmetric orbital forcings provides a hint, however.

Warmings proceed more rapidly than coolings, presumably because the growth of an ice sheet is limited by the rate of snowfall in a cold place, while multiple amplifying feedbacks can speed the wet process of ice sheet disintegration, once it begins in earnest.

I realized early in this decade that there was a growing danger of pushing the climate system to a point such that future disasters might occur out of our control. The concepts are not difficult. A look at just two phenomena, inertia and feedbacks, is enough to yield the conclusion: We really do have a planet in peril.

Three big sources of inertia affect global warming and its consequences: the ocean, the ice sheets, and world energy systems. The ocean is, on average, about two and a half miles deep. It takes the ocean a long time—centuries—to fully warm up in response to human-made greenhouse gases. So even if we stabilize atmospheric composition at today's levels, the planet will still continue to heat up, because the ocean will continue to warm. If the ocean were the only source of inertia, additional warming over the course of this century—with no additional gases—would be a few tenths of a degree Celsius.

The nature of the second source of inertia, ice sheet inertia, is—in one key sense—almost opposite that of the ocean. Despite the dynamical tricks the ocean can play, its thermal inertia effect is pretty straightforward. Ocean surface temperature, the quantity that most affects global climate, achieves half of its equilibrium (long-term) response to a forcing within a few decades. Yes, it takes many more decades, even centuries, for the full response, but the ocean has already achieved about half or more of its full response to greenhouse gases added to the air in the past century.

Ice sheet response to global warming is quite the contrary. Ice sheet size changes little at first, and thus sea level changes only slowly. As the planet gets warmer, the area on the ice sheet with summer melt increases. And as the ocean warms, ice "shelves"— tongues of the ice sheet that reach out into the ocean and are grounded on the ocean floor—also begin to melt. As ice shelves disappear and the ice sheet is "softened up" by surface warming and meltwater, movement of ice and discharge of giant icebergs via ice "streams" become more rapid, leading to the possibility that large portions of the ice sheet will collapse.

If we continue burning fossil fuels at current rates, ice sheet collapse and sea level rise of at least several meters is a dead certainty. We know this from paleoclimate records showing how large the ice sheets were as a function of global temperature. The only question is how fast ice sheet disintegration will occur.

Once ice sheets begin to collapse, sea level can rise rapidly. For example, about 14,000 years ago, as Earth emerged from the last ice age and became warmer, sea level rose at an average rate of 1 meter every 20 or 25 years, a rate that continued for several centuries. The danger today is that we may allow ocean warming and "softening up" of ice sheets to reach a point such that the dynamical process of collapse takes over. And then it would be too late—we cannot tie a rope or build a wall around a mile-thick ice sheet.

The third source of inertia is our fossil-fuel-based energy system. The transitions from wood to coal to oil to gas each required several decades—and recently, as oil and gas supplies tightened, we have begun moving back toward more coal use. Indeed, coal is again the largest source of carbon dioxide emissions.

The upshot regarding energy system inertia is this: Humanity today is heavily dependent on fossil fuels—coal, oil, and gas—for most of our energy. When we realize that it is necessary to phase out fossil fuels, that transition will not be quick—it will take at least several decades to replace our enormous fossil fuel infrastructure. In the meantime more greenhouse gas emissions and more climate change will be occurring.

Climate feedbacks interact with inertia. Feedbacks (as discussed in chapter 3) are responses to climate change that can either amplify or diminish the climate change. There is no inherent reason for our climate to be dominated by amplifying feedbacks. Indeed, on very long time scales important diminishing feedbacks come into play (see chapters 8 and 10).

However, it turns out that amplifying feedbacks are dominant on time scales from decades to hundreds of thousands of years. Water (including water vapor, ice, and snow) plays a big role. A colder planet has a brighter surface and absorbs less sunlight, mainly because of the high reflectivity of ice and snow surfaces. A warmer planet has more greenhouse gases in the air, especially water vapor, as well as darker vegetated land areas. Dominance of these two amplifying feedbacks, the planet's surface reflectivity and the amount of greenhouse gases in the air, is the reason climate whipsawed between glacial and interglacial states in response to small insolation changes caused by slight perturbations of Earth's orbit.

Amplifying feedbacks that were expected to occur only slowly have begun to come into play in the past few years. These feedbacks include significant reduction in ice sheets, release of greenhouse gases from melting permafrost and Arctic continental shelves, and movement of climatic zones with resulting changes in vegetation distributions. These feedbacks were not incorporated in most climate simulations, such as those of the Intergovernmental Panel on Climate Change (IPCC). Yet these "slow" feedbacks are already beginning to emerge in the real world.

Rats! That is a problem. Climate inertia causes more warming to be in the pipeline. Feedbacks will amplify that warming. So "inertia" was a Trojan horse—it only seemed like a friend. It lulled us to sleep, and we did not see what was happening. Now we have a situation with big impacts on the horizon—possibly including ice sheet collapse, ecosystem collapse, and species extinction, the dangers of which I will discuss later.

What to do? If we run around as if our hair is on fire, flapping our arms, people will not take us seriously. Besides, we are not in a hopeless situation. Rational, feasible actions could avert disastrous consequences, if the actions are prompt and strategic. Feedbacks work in both directions—if a forcing is negative, amplifying feedbacks will increase the cooling effect.

If we wish to stabilize Earth's climate, we do not need to return its atmospheric composition to preindustrial levels. What we must do, to first order, is reduce the planet's energy imbalance to near zero. Of course, the climate then would be stabilized at its current state, not at its preindustrial state. Climate may need to be a tad cooler than today, if, for example, we want ice sheets to be stable. That may require a slight additional adjustment of the human-made climate forcing. But let's not get ahead of the story.

In December 2001 I received a letter from the editor in chief of *Scientific American*, John Rennie, inviting me to write a 3,500-word article on global warming. It was an opportunity to describe the climate crisis to a broader audience, and I was eager to convey several points.

First, I wanted to make clear the danger that business-as-usual emissions would lead to eventual ice sheet disintegration and large sea level rise. Second, in contrast to a 2001 IPCC report implying that global warming of about 3 degrees Celsius (above the 1990 level) was needed to reach the dangerous level, it seemed clear to me that 3-degree global warming, or even 2-degree, was a recipe for global disaster. Third, scientists needed to define scenarios that would keep global warming within tolerable limits. Otherwise we aid and abet government energy departments that seem to be working hand in glove with the fossil fuel industry, accepting as a god-given fact that humanity will proceed to burn all fossil fuels.

In 2002 I published six papers, organized a major workshop at the East-West Center in Hawaii, and published a report on that workshop. But, for me, writing 3,500 words for a public audience was harder. I kept plugging away at it, and by late 2002 I thought I had something good. However, when I submitted the article, titled "Can We Defuse the Global Warming Time Bomb?" to *Scientific American*, the editor assigned to work with me began making extensive changes, to my great consternation.

True, my article was not exactly what was requested, but in my opinion it was better. The invitation letter had asked me to make the case for global warming, its causes, and consequences, "citing how researchers are becoming more comfortable about forecasting regional climate change and sea level rise based on ever-improving climate models." Well, I had been working with climate models for decades, and I knew that some of the most recent models predicted ice sheets would *grow* with global warming, causing sea level to fall, defying common sense and empirical evidence. Models are no better than the representations of processes that are put into them—and even if you put in a good description of a process, another deficient part of the model may completely screw up the result. In the case of ice sheets, some of the most important processes were not even included in the climate models.

I prefer to start with paleoclimate, the lessons of history, which provide our best measure of how Earth responds to changing boundary conditions or forcings. Second, as a measure of how rapidly climate can change, we need to look at what is happening now—observations of the ongoing climate response to fast-changing human and natural forcings. Climate models come third. Models aid interpretation of past climate, and they are needed to project future changes. So models are valuable, but only when used with knowledge of their capabilities and limitations.

What upset me most was the insertion, by the editor, of the approach and perspectives of IPCC. My aim was to give a *different* perspective on climate change. I was implicitly critical of IPCC— its minimization of likely sea level rise under business-as-usual forcing, its high estimate for the dangerous level of global warming, and the absence of any effort to define scenarios that would avert the dangers of our current energy policies. The editor asserted that he was only clarifying the story, but in the process he was inserting his and IPCC's perspective.

Further communication with the editor revealed that even more editing was planned to make the article better correspond to the magazine's concept. In anger, I withdrew the article, saying that it could be published only on the condition that not a single word was changed. Of course that was unacceptable—and unfortunately, with the paper's extensive criticisms of IPCC, there was no realistic chance of publishing it in a regular scientific journal—most of the likely referees for the paper were contributing authors of IPCC. I decided that I would prefer to publish it in the gray literature, as a report or on the Web. It was disappointing, yet a relief—I escaped the 3,500-word constraint of *Scientific American* and could make the story more quantitative and complete.

So I prepared a much longer document, still using the title "Can We Defuse the Global Warming Time Bomb?" This paper included most of the charts that I used in my presentation to the White House Council of Environmental Quality on June 12, 2003, and served as a record of that presentation.

The principal contribution of this document was an attempt to define the "dangerous" level of global warming. I concluded that ice sheets were a critical issue—and the more I looked at paleoclimate data, the more I realized how sensitive ice sheets were to even small global mean warming. My inference was that global warming (above that in 2000) should be kept to less than 1 degree Celsius. That limit implied that CO_2 would need to peak at about 450 parts per million—or perhaps 475 ppm, if substantial but plausible reductions of non-CO_2 forcings were achieved. (As mentioned earlier, I've since revised that target limit downward, to 350 ppm.)

In July 2003 I received a request from the Director's Office at the NASA Goddard Space Flight Center, in Greenbelt, Maryland, to give a presentation to NASA administrator Sean O'Keefe. The director of earth sciences at Goddard, Franco Einaudi, told me that it would be the first earth science presentation to O'Keefe, who had been administrator for more than a year and had already visited Goddard once. On that earlier visit, he declined an offer to attend science presentations, instead choosing to visit the "visualization" laboratory, where he watched a film clip prepared for television viewers of the Winter Olympics; the clip started with a view of the whole Earth from space and then steadily zoomed in to show the Olympics site in Salt Lake City.

The Goddard director offered, as an alternative to my talk, to provide a presentation on Goddard's Earth observation satellites, but the Administrator's Office responded that O'Keefe wanted to hear from me about black carbon. I decided to use the contents of my "Time Bomb" document—it included black carbon but in the context of the actions needed to stabilize climate.

I sent a copy of "Can We Defuse the Global Warming Time Bomb?" to Ghassem Asrar, NASA's associate administrator of earth sciences, who would accompany O'Keefe on his visit to Goddard. As we were assembling in the Goddard director's office for my presentation, Dr. Asrar showed me the version of the paper that he had given to O'Keefe—Asrar had changed the title to something about "climate change" to make it less "incendiary."

Sean O'Keefe was a friend and protégé of Vice President Dick Cheney. An accountant by training, O'Keefe had worked in the Department of Defense and on the staff of the Senate Appropriations Committee Subcommittee on Defense. In March 1989, two months after Cheney became secretary of defense, he had O'Keefe successfully nominated for Defense Department comptroller and chief financial officer. Mark Bowen, in *Censoring Science*, notes that O'Keefe was "openly and unapologetically partisan. As one senior insider at the agency [NASA] puts it, 'In came Sean, and then it became very clear that NASA belonged to Sean, who belonged to Cheney.'" O'Keefe is the only NASA administrator who was not trained in science and engineering.

O'Keefe, a pleasant, soft-spoken person, listened quietly until I showed a photograph of a raging stream of meltwater on Greenland

FIGURE 6. *A stream of snowmelt cascades down a moulin near Ilulissat, Greenland, in 2008. A moulin is a near-vertical shaft worn in the ice sheet by the meltwater. (Photograph courtesy of Konrad Steffen.)*

plummeting into a moulin. The chart's title was "What Determines 'Dangerous Anthropogenic Interference'?" I took the latter phrase from the 1992 Framework Convention on Climate Change treaty, signed by practically all nations of the world, including the United States. The countries agreed to take steps to keep greenhouse gases at a level that would avert dangerous climate change, with the steps to be defined in binding protocols. The main conclusion of my "Time Bomb" paper was that the stability of the Antarctic and Greenland ice sheets would surely set a low limit on permissible global warming and thus a low limit on greenhouse gases.

O'Keefe interrupted me to say that he did not think I should use the "dangerous" phrase, because we did not understand climate well enough to say what constituted danger. I probably *should* have disputed his admonition—our ignorance of what constitutes danger is actually a reason to focus on that topic. But nobody wanted to see a disagreement with the administrator, and I could readily admit that our understanding of ice sheet behavior was rudimentary, so the ensuing discussion was brief and not heated.

O'Keefe did not "order" me to never again use the "dangerous" phrase; it was only polite but unequivocal advice. Franco Einaudi remembers O'Keefe's admonition as simply a mild rebuke. Yet it was O'Keefe's only interjection during my entire talk, it was crystal clear, and he was the administrator.

During the return trip to New York I at first felt bad about the exchange—and my failure to take issue with the administrator's advice. Of course, arguing the point would not have altered his opinion. But this matter raised a question: Given the fundamental nature of the "dangerous" phrase in the Framework Convention, why was there not greater explicit attention to it in the scientific literature? I decided that I would try to draw more attention to this "dangerous" issue.

I added some clarification and another figure to the "Time Bomb" document and sent the resulting version to be posted on the rather obscure Web site naturalSCIENCE. At about the same time, I received a proposal from the *Scientific American* editor: They were willing to publish this entire "Time Bomb" document on their Web site as well as a condensed version in *Scientific American*—with no changes, only condensation.

This second go-round with *Scientific American* worked out better. The editor said that he was busy on another project, so an alternate editor would work with me on extracting the condensation. I guessed that the change of editors was intended to minimize the chance that I would fly off the handle again.

The word limit for my *Scientific American* article was increased to 4,500 words, but I knew I would need still more space to adequately discuss why I thought ice sheets are closer to dangerous disintegration than IPCC assumed. An opportunity to supplement the *Scientific American* discussion was provided by the coincidence of an invitation from Steve Schneider, editor of the journal *Climatic Change*, to write an editorial essay, and a request from the State Department to attend a European Union–United States climate workshop in Bologna, Italy. I used my spare time on that trip to write the essay "A Slippery Slope: How Much Global Warming Constitutes 'Dangerous Anthropogenic Interference.'"

People ask me why the "Slippery Slope" paper acknowledges Harlan Watson for facilitating the paper. It is because he used his gold card to get me into his airline's first-class lounge in Munich, where I spent a six-hour layover writing half the paper. Watson was heavily criticized for being the face of the Bush administration's rejection of the Kyoto Protocol. But I will show quantitatively in chapter 9 that Watson's assertion that the protocol "was more about being seen to agree than about actual action" was dead on the mark. And I will argue that there is a great danger that our governments will follow a similar ineffectual path in the next international agreement, unless the public places strong pressure on them.

Publication of "Slippery Slope" took a year. The editor decided, because the paper challenged IPCC and conventional wisdom, that there should be a commentary on it, which he obtained from Michael Oppenheimer and Richard Alley. Then my essay in *Climatic Change* took another half year to be published, because of old-fashioned typesetting procedures.

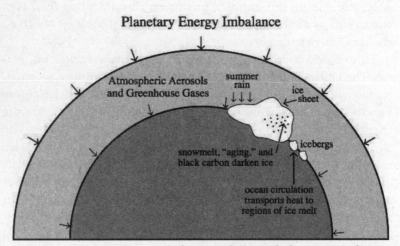

Planetary Energy Imbalance

FIGURE 7. *Earth's energy imbalance is deposited almost entirely into the ocean, where it contributes to iceberg and ice shelf melting. After ice sheet disintegration begins, a substantial fraction of the energy imbalance may go into melting ice. (Figure from Hansen, "A Slippery Slope." See sources.)*

What I was trying to address in this essay were IPCC estimates that if greenhouse gas emissions increased on a business-as-usual path, sea level might rise about a foot (30 centimeters) or perhaps a foot and a half in a century. Such a sea level rise would be more than a nuisance but hardly a disastrous global alteration of shorelines. The IPCC picture seemed to allow plenty of time to study the matter more carefully, and perhaps agree on ways to adapt to such changes. Sea level rise, in the panel's estimates, would be due to the melting of mountain glaciers and the expansion of ocean water as it became warmer.

But IPCC sea level change estimates did not include any contribution from Greenland or Antarctica. Its rationale: Global warming might speed melting at the edges of ice sheets, but a warmer atmosphere would also increase winter snowfall, which would thus make the central part of the ice sheets thicker. Indeed, as I wrote the "Slippery Slope" paper in 2003, the most recent global climate model results—from one of the best models in the world, with the highest resolution—were published in the *Journal of Geophysical Research*. They concluded that the ice sheets would grow as the world became warmer, thus tending to make sea level fall.

Hmm, is something wrong with that picture? As the planet warms, ice sheets get bigger? Actually, that is conceivable, for a limited period. Rapid atmospheric warming could cause a prompt snowfall increase that exceeds increased ice loss at the ice sheet periphery, if changes of ice sheet dynamics begin slowly. The problem is that most existing climate models pretty much *assume* that result will occur, by treating ice sheets as if they were giant rigid ice cubes that melt only slowly. Models, at best, produce answers consistent with the assumptions put into them.

The diagram I included in the essay (similar to figure 7) was intended to aid discussion of processes that are not adequately represented in global climate models. It also provides a different way to think about the sea level problem. It focuses on the planet's energy imbalance—where the excess energy goes and why it is important for ice sheets.

Earth's energy imbalance is tiny. In 2003 I estimated that it was between 0.5 and 1 watt per square meter. The latest data suggest about 0.5 watt, averaged over several years and averaged over the planet. Yet this small energy imbalance is the most important number characterizing the state of our climate. It defines how much more global warming is "in the pipeline" without further change of atmospheric composition, and it tells us how much we must alter human-made climate forcings if we want to restore the planet's energy balance and thus, to first order, stabilize climate—topics we will dig into in later chapters.

For the moment, it's important to note that climate models have a hard time estimating the imbalance because it depends on the net climate forcing, and the big aerosol forcing is unmeasured. Thus our best measure of Earth's energy imbalance comes from toting up observed changes of energy in its reservoirs—the ocean, atmosphere, land, and ice. It turns out that the lion's share of the excess incoming energy, about 90 percent, goes into the ocean.

Let's first make a calculation along the lines that IPCC assumes, namely with the ice sheet melting as if it were a giant ice cube. In that case, how much of the ice would melt due to human-made heating, which we take as, say, 1 watt per square meter? It is an easy calculation—most of the energy is needed for the phase change from ice to liquid. That requires 80 calories of energy for each gram of ice (if you are a youngster in a physics class today, rather than an old guy like me, you would say that it requires about 335 joules of energy—either unit is okay). If the average melt season is four months long and covers one third of Greenland, then the extra melt due to 1 watt of heating is about 20 cubic kilometers of water—enough to raise global sea level 0.05 millimeter (5 millimeters in a century). Hmm—not very much. Moreover, climate models find that global warming increases winter snowfall by more than that, so the net effect would be a sea level decrease as Earth becomes warmer.

How do we know this picture is wrong? Earth's paleoclimate history shows the contrary: As Earth gets warmer, ice sheets get smaller and sea level rises. Indeed, sea level sometimes rises as much as several meters per century. Where does the energy to melt the ice so fast come from? The explanation surely must involve the huge reservoir of energy provided by the ocean and it must include ice sheet dynamics.

Let's consider the amount of energy being soaked up by the ocean. As a round number, which we can scale later to any fraction that we want, let's say that Earth's energy imbalance is 1 watt per square meter, with 90 percent of the energy going into the ocean. If all that energy were used to melt ice, sea level would rise 10 centimeters per year (or 10 meters, about 33 feet, per century). For an Earth energy imbalance of 0.5 watt, these numbers would be half as large.

A 1- to 2-meter sea level rise would be disastrous for hundreds of millions of people. So if even a fraction of the excess energy going into the ocean finds its way to the ice sheets, we are in trouble. I argue that Earth's history demonstrates that there are efficient ways to transfer energy between the ocean and ice. Furthermore, observations suggest that these mechanisms are beginning to come into play today.

One mechanism is the melting of ice shelves by the warming ocean. Ice shelves—tongues of the ice sheets that extend into the ocean, usually at least partially grounded on the ocean floor—buttress the ice sheet, limiting the rate at which ice is discharged to the ocean. This buttressing opposes the natural "plastic" flow of ice toward the ocean, which is driven by the weight of the ice sheet as snowfall piles up on its interior. If a warming ocean melts ice shelves, the ice "streams" coming from the ice sheet, which discharge giant icebergs into the ocean, begin to move more rapidly, discharging more ice. It is somewhat analogous to pulling the cork from a wine bottle—removing the impediment allows rapid flow.

The West Antarctic ice sheet is especially vulnerable to removal of its ice shelves, because much of that ice sheet rests on bedrock several hundred meters below sea level. Loss of the entire West Antarctic ice sheet would raise sea level 6 to 7 meters (20 to 25 feet) and eventually open a path to the ocean for part of the much larger East Antarctic ice sheet. Once the ice sheets' collapse begins, global coastal devastations and their economic reverberations may make it impractical for humanity to take actions to rapidly reverse climate forcings. Thus if we trigger the collapse of the West Antarctic ice sheet, sea level rise may continue to even much higher levels via contributions from the Greenland and East Antarctica ice sheets.

Most of the Greenland ice sheet sits on bedrock above sea level, but some of Greenland's major ice streams are in fjords with the bedrock well below sea level. The termini of these ice streams are retreating into the ice sheet as warming ocean melts the ice front. If the warming continues and termini are pushed farther back into the ice sheet, walls of ice sheet on both sides of fjords may begin to collapse, increasing the rush of giant icebergs to the ocean.

Disappearing ice shelves, ice stream dynamics, and iceberg melting were not included in global climate models used for IPCC studies. This failure to take into account the increased discharge of icebergs to the ocean, where they melt much more rapidly than they would if they had remained as an ice block on land, probably explains the models' inability to predict realistic sea level change. It is not necessary to move excess heat from the ocean to the ice sheets in order for ice sheets to shrink. Rather the mountain can come to Muhammad: Chunks of the ice sheet (icebergs) are dispersed over a broad area, where they melt by drawing heat from ocean water.

Melting ice shelves is the critical mechanism in initiating ice sheet collapse. However, other contributing factors and feedbacks speed ice sheet disintegration. As the atmosphere becomes warmer, "aging" of snow accelerates—that is the process in which snow crystals vaporize on a microscopic scale and re-form into larger, darker crystals, which absorb more sunlight. Also, snowmelt begins earlier in the spring, causing the ice sheet to also become darker and absorb more sunlight. Human-made black soot aerosols, which are now deposited in measurable quantities on the Greenland ice sheet, contribute to this process as well. And as ice sheet mass loss becomes substantial, the ice sheet surface sinks to a lower level, where the temperature is warmer, which is another amplifying feedback.

Given these amplifying feedbacks, it is no wonder that the glacial-interglacial climate cycles depicted in figure 3 (page 37) are asymmetric, with the wet process of ice sheet disintegration proceeding much more rapidly than ice sheet growth. Sea level rise at a rate of a few meters per century is not uncommon in the paleoclimate record. Instead, it is the stability of sea level for the past 7,000 years that is unusual. Earth in recent millennia was warm enough to prevent an ice sheet from forming in Canada but cool enough to keep the Greenland and Antarctic ice sheets stable. Also, any tendency for continued ice sheet mass loss after the demise of the large Laurentide (North American) ice sheet was opposed by the slight global cooling trend since peak early Holocene temperatures (6,000 to 10,000 years ago).

As mentioned earlier, the sea level stability of the past 7,000 years probably contributed to the development of civilization, because stable sea level led to high biologic productivity and thus ample amounts of fish in coastal areas. With the exception of Jericho, the first cities that developed on several continents 5,000 to 7,000 years ago were all coastal cities. Even today a large portion of the world's cities are located along the coasts; more than a billion people live within a 25-meter elevation of sea level.

If ice sheets begin to disintegrate, there will not be a new stable sea level on any foreseeable time scale. Instead, we will have created a situation with continual change, with intermittent calamities at thousands of cities around the world. Because the ocean and ice sheets each have response times of at least centuries, change will continue for as many generations as we care to think about. Change will not be smooth and uniform. Instead, local catastrophes will occur in association with regional storms. Given the enormous infrastructure and historical treasures in our coastal cities, it borders on insanity to suggest that humans should work to "adapt" to climate change, as opposed to taking actions needed to stabilize climate.

Would coastal cities be rebuilt, given the knowledge that sea level will continue to rise? It is hard to imagine that humanity would decide to abandon coastlines—although look at New Orleans. But where would people in low-lying regions such as Bangladesh migrate to? Global chaos will be difficult to avoid if we allow the ice sheets to become unstable.

Was sea level stable during prior interglacial periods, some of which were warmer than the Holocene? Bill Thompson of the Woods Hole Oceanographic Institute deduced, from heights of ancient coral reefs on eroded shorelines, that sea level fluctuated several meters during the last interglacial period, about 120,000 years ago. Geologist Paul Hearty used another indicator of past sea level, wave-formed shoreline terraces, to draw similar conclusions. Recently Paul Blanchon and colleagues at the University of Mexico presented evidence that a 2- to 3-meter sea level rise probably occurred in a period of 50 years or less during that interglacial period. Such a rapid change would imply ice sheet collapse, most likely on West Antarctica.

Sea level changes to heights at least several meters greater than today's level occurred in interglacial periods that were at most 1 to 2 degrees Celsius warmer than today. As this knowledge was developing and becoming more convincing to me, I argued that we must keep additional global warming to less than 1 degree Celsius, much below the 2- to 3-degree "dangerous" level that IPCC suggested with their well-known "burning embers" diagram—used to indicate the probablility of danger as a function of global warming, it begins to glow red, for danger, only when global warming exceeds 2 to 3 degrees Celsius. It was this rationale that led me to argue for a maximum CO_2 level of about 450 ppm, as discussed in the "Time Bomb" and other papers.

The paleoclimate sea level data were complemented by disturbing data on ongoing changes in polar regions. Eric Rignot of the Jet Propulsion Laboratory reported that most of the ice shelves around Antarctica were melting from below at a rate of several meters per year. This melting clearly was due to warming ocean waters, although there was no proof that the warming was human-caused.

Melting on the Greenland ice sheet also increased. Summer melting fluctuates year to year, depending on the weather, but there was a clear long-term increase of melt area. In fact the area with melting has almost doubled since the beginning of satellite measurements in the late 1970s. Estimates of ice sheet mass balance, gains from snowfall and loss from melting, show that both the Greenland and West Antarctica ice sheets are beginning to lose mass. Sea level is observed to be rising at a rate of more than 3 centimeters per decade. That is a rate of about a foot per century, twice as large as the rate of sea level rise in the twentieth century.

This information was becoming clear in early 2004 when I received the proofs for my "Time Bomb" paper, scheduled for publication in March of that year, which included a Greenland moulin photo (similar to figure 6). I sent a note to glaciologist Jay Zwally, asking if I would be crucified if I included this caption: "On a slippery slope to Hell, a stream of snowmelt cascades down a moulin on the Greenland ice sheet. The moulin, a near-vertical shaft worn in the ice by surface water, carries water to the base of the ice sheet. There the water is a lubricating fluid that speeds motion and disintegration of the ice sheet. Ice sheet growth is a slow process, inherently limited by the snowfall rate, but disintegration is a wet process, spurred by feedbacks, and once well under way it can be explosively rapid."

Zwally replied, "Well, you have been crucified before, and March is the right time of year for that, but I would delete 'to Hell' and 'explosively.'" I thought immediately of a fellow who had gone over Niagara Falls a year earlier without a barrel (and lived to tell about it). That would seem like a joy ride compared with slipping on the banks of the rushing meltwater, clawing desperately in the freezing water before being hurtled down the moulin more than a kilometer, eventually being crushed by the giant, grinding glacier. But I was using "slippery slope" mainly as a metaphor for the danger posed by global warming. So I changed "Hell" to "disaster."

What about "explosively"? Paleoclimate sea level increase as great as 1 meter in 20 years is 15,000 cubic kilometers of water per year. Ice sheet disintegration at even a fraction of that rate would seem pretty explosive to most people.

That photograph caption first caused me to think about "scientific reticence." Reticence leaped to mind again a year later as I was being grilled by a nasty lawyer for the plaintiff in an automobile manufacturers versus the California Air Resources Board lawsuit. He demanded that I name a glaciologist who agreed, on the record, with my assertion that business-as-usual greenhouse gas emissions would likely cause a sea level rise of at least a meter in a century: "Name one!"

I could not, instantly. I was dismayed, because I sensed a deep concern among relevant scientists about likely consequences of continued emissions growth. I remembered a field glaciologist saying, in reference to a moulin, "The whole damned ice sheet is going to go down that hole!"

Scientific reticence, in some cases, may hinder communication with the public. Reticence may be a consequence of the scientific method—success in science depends on continual objective skepticism. Caution has its merits, but we may live to rue our reticence if it serves to help lock in future disasters.

I could not use the lame excuse of "scientific reticence" in a face-off with the automakers' lawyer. But I knew that scientific reticence was a real phenomenon, and I eventually wrote a paper on the topic ("Scientific Reticence and Sea Level Rise," published in *Environmental Research Letters* in May 2007). One factor in reticence may be "behavioral discounting"—concern about the danger of being accused of "crying wolf" is more immediate than concern about the danger of "fiddling while Rome burns." In other words, a preference for immediate, over delayed, rewards may contribute to irrational reticence even among rational scientists.

"Crying wolf" can affect funding. I call it the "John Mercer effect." Mercer warned, in the late 1970s, that burning fossil fuels may lead to disintegration of the West Antarctic ice sheet, with a sea level rise of several meters. I noticed at the time that the Department of Energy treated the scientists who suggested that Mercer was alarmist as more authoritative.

Drawing attention to the dangers of climate change may or may not have helped overall earth science funding, but it surely did not help individuals like Mercer who stuck their necks out. I could vouch for that from my own experience. After I published a paper in *Science* in 1981 that described likely climate effects of fossil fuel use, the Department of Energy reversed a decision to fund my research, specifically citing criticisms of that paper as being alarmist.

Not until the 2007 IPCC report came out were a few scientists spurred into indignation at the panel's failure to draw adequate attention to the danger of sea level rise. In the 2007 report, IPCC actually lowered its estimated sea level rise from the previous report, as it still neglected mass loss by the Antarctica and Greenland ice sheets while reducing its calculated thermal expansion of ocean water. Stefan Rahmstorf, in a well-publicized paper published in *Science*, pointed out that if sea level rise was simply assumed to be proportional to global temperature change, the sea level rise of the past century implied that business-as-usual global warming would yield sea level rise of about a meter within a century. Rahmstorf's paper was a big help in waking up policy makers to the threat of sea level rise.

Even so, I had the feeling that the scientific community was just nudging public knowledge in the direction where many scientists suspected that the answer would lie. It reminded me of Feynman's story about Robert Millikan's oil drop experiment. Millikan's experiment was famous because it showed that electric charge was quantized to exact multiples of an elementary electron charge. But the value Millikan derived for the electron charge was not quite right. Subsequent experimenters began to report values a bit larger, then still larger, and so on—until eventually they converged on what we now know is an accurate value. The scientists always had enough uncertain parameters in their experimental setup that they could rationalize choices yielding a result not too different from Millikan's. This history suggested a reticence of scientists to report a result that differed too much from the one established by the great Millikan—at least not in a single step—until other scientists gave them more courage.

Feynman liked to needle his experimentalist colleagues about this. Their exposed reticence amused Feynman and embarrassed them, but it was not a big problem for the world. We cannot say the same for reticence about sea level rise, if that reticence delays actions needed to avert a disaster.

In a nutshell, a problem has emerged. Climate inertia and climate amplifying feedbacks, as humans rapidly increase atmospheric greenhouse gases, spell danger for future generations—big danger. Yet the public is largely unaware of an impending crisis. The obliviousness of the public is not surprising—global warming, as yet, is slight compared to day-to-day weather fluctuations. How in the world can a situation like this be communicated credibly?

CHAPTER 6

The Faustian Bargain: Humanity's Own Trap

On June 3, 2004, I received a letter from Frank Loy, the chief climate negotiator for the United States during the Clinton-Gore administration. The letter said that I had been selected by the board of directors of Environment 2004, including Bruce Babbitt and Carol Browner, to receive an award for my climate research and advocacy. I was also invited to join the group's leadership council and be a featured speaker at a reception in New York with about seventy-five guests. It seemed to be primarily a fund-raising event.

Environment 2004 was, according to its mission statement, "dedicated to assuring the defeat of President George W. Bush and his allies by highlighting the environmental stakes in the next election." It planned to "make a difference in November by swinging a significant number of voters in a handful of hotly contested swing states that will likely decide the Presidential and a few statewide elections."

Perhaps this was a chance to clarify the slippery slope that humanity was treading. But I was dubious about both attending the reception and accepting the award. As a registered Independent voter, I prefer not to be tied to either major party, so I did not respond right away.

After my 1988 and 1989 congressional testimony, I had declined most requests for talks and interviews, especially for television. Interviews are a time-consuming distraction from research. Besides, even if I had spent time preparing, I still felt awkward and inarticulate. My fear of speaking was not as bad as it had been earlier in my career, when my brain seemed to freeze up before an audience.

Once, at a Pioneer Venus mission meeting in the 1970s, when I went to show a viewgraph, I could not think, so I just went back to my seat—which was very embarrassing. But my colleagues Michael Oppenheimer and Steve Schneider, who gave excellent interviews, were willing to take referrals—so beginning in 1989 I directed global warming interview requests to them.

That worked fine for fifteen years, allowing me to do hands-on science, despite administrative and fund-raising responsibilities. I liked to work with a small number of people on problems where we could experience what Richard Feynman called "the pleasure of finding things out." Mainly I worked with Reto Ruedy, a mathematician and programmer for our global climate model, and Makiko Sato, a physicist who did her Ph.D. research on Jupiter's atmosphere. We developed a simplified version of the Goddard Institute's climate model that was fast enough to run hundreds of experiments, and we developed data sets for various observed climate parameters. For example, when the Mount Pinatubo volcano in the Philippines erupted in 1991, we used our model and a data set for stratospheric aerosols to study the effect of the volcano on global temperature. Our prediction of global cooling over a two-year period turned out to be accurate, increasing our confidence that we understood the effect of climate forcings on global temperature.

Frank Loy's letter ended that period of sticking strictly to science. It forced me to think about the gap between what was understood about global warming by the relevant scientific community and what was known about global warming by the people who need to know, the public. Scientists who studied ice sheet stability were concerned that the planet was headed toward disastrous consequences, but they seemed reticent to speak out. I had concluded that further global warming (above that in 2000) must be kept to less than about 1 degree Celsius to avert disaster. But both Oppenheimer and Schneider published papers suggesting that the dangerous level of warming was much more distant, about 2 to 3 degrees Celsius. I couldn't continue to defer interview requests to them; I would need to speak for myself.

It was also becoming apparent that NASA, consistent with Administrator Sean O'Keefe's admonishment not to talk about "dangerous human-made interference," seemed reluctant to publicize papers that drew attention to climate concerns. And the Bush administration was not willing to revise its climate policies in the face of new data, contrary to the president's 2001 Rose Garden declaration. But who was to blame? Future generations might look back and say, "How could they have been so stupid? Why didn't they do anything?" "They" would include scientists who did not adequately communicate the danger.

Those were the thoughts that led me to conclude, as I have often repeated, that I did not want my grandchildren to say, "Opa understood what was happening, but he didn't make it clear." (Actually, I am known as "Bopa" in our family—that is how Sophie, our first grandchild, pronounced *opa*, Dutch for "grandpa." We liked her pronunciation and chose to let it stick.) Thus was born my decision to give one public talk in which I would lay out the story as clearly as I could. Then I would get back to science.

I decided to decline the award and speaking invitation from Loy but ask for his help in finding a venue for a public talk, to be given as a private citizen. The talk would be consistent with his aim of making climate a campaign issue. But I wanted a nonpartisan venue, even though, at least implicitly, I would be critical of the Bush administration's lack of action.

I did not want to call Loy from my office—not because I believed the assertions of a man who said he was certain that my phone was tapped, which seemed unlikely, but because the talk needed to occur independent of my government job. On my next trip to Washington, I called Loy from my cell phone outside Union Station, and he agreed immediately that such a talk would be useful. He suggested that it be given in Washington, hosted by Resources for the Future, on whose board of directors he served. Loy said he would make sure that the talk received extensive media coverage.

Thus began a four-month period of unusual pressure and intensive work, even by my standards. Unlike my 1988 testimony before Congress, which was based on a scientific paper that had been accepted for publication, this talk would be based on papers still in preparation. If those papers were not ready by the time of the talk, I could be open to criticism by others in the scientific community, and the talk could backfire.

My "Slippery Slope" paper was in press, but it was an opinion piece. My paper titled "Efficacy of Climate Forcings," which provided a comprehensive comparison of a broad range of climate forcing mechanisms, had been in preparation for more than a year. Other papers in preparation investigated Earth's current energy imbalance and compared climate simulations for the past century with observational data on climate change.

It almost always takes longer to complete a paper than early estimates allow. Jim Pollack and I once joked that there was a missing factor of pi (about 3) in such estimates. But I could not afford such delay. I began waking up in the middle of the night, working for a few hours to make use of that time, and then would take a dose of cold and flu medication to help me get back to sleep.

Summer came and went too rapidly. Time was running out if the talk was to be relevant to the 2004 election. Even though the papers were not quite ready, I called the contact person at Resources for the Future, asking for the talk to be scheduled in early October. There was hesitation, a concern about possible political ramifications that had not been expressed before. It required checking with management.

I got the response on October 4 by e-mail: Resources for the Future had placed the seminar series "on hiatus until early 2005 as part of a reevaluation of our outreach and public education agenda." The contact person wrote, "We felt this was a good time to review this part of our work, given the national elections that seem to be sucking the policy air out of the room, and the fact that RFF is launching a major policy book in mid-November that is going to occupy much of our time and attention for the rest of the year."

I felt a sudden deflation, with mixed emotions. It no longer mattered that my papers were unfinished. I could focus on just the science again. But the intense effort of the preceding months was for naught. There would be no chance to better inform the public, prior to the election, about the urgent need to change course on climate policy.

The next day I received a message from Leslie McCarthy, a public affairs employee at the Goddard Institute for Space Studies (GISS) in New York. She included an e-mail message from Rob Gutro, a public affairs employee at the Goddard Space Flight Center, in which Rob informed her about the status of a press release that had been held up at NASA headquarters for a month.

Gutro's e-mail read: "According to HQ, there's a new review process that has totally gridlocked all earth science press releases relating to climate or climate change. According to HQ Public Affairs, 2 political appointees, Ghassem and the White House are now reviewing all climate related press releases . . . thus, the 4+ week review time for Drew's press release that was slated for issue on Sept. 27th. We're still waiting to get the release back from the WH. We'll let you know when it happens."

The specific paper referred to was by GISS scientist Drew Shindell. Ghassem Asrar was NASA's associate administrator for earth sciences. As described in Mark Bowen's *Censoring Science*, the Office of Public Affairs at NASA headquarters had been taken over by political appointees of the Bush administration. With Gutro's revelation that press releases were being spirited to the White House, where they were held up and edited, it seemed to me that NASA's Office of Public Affairs had become its Office of Propaganda.

I could not sleep that night. I got up and wrote a letter to my former mentor at the University of Iowa, Professor James Van Allen— it was two pages, single-spaced with small margins, and it took about half the night to write. I summarized the climate story, my meetings with the cabinet-level Task Force, and the new revelation that NASA headquarters was cooperating in selective reporting of results—it seemed that by "sound science" the administration meant science that gave the "right" predetermined answer. I described the debacle with Resources for the Future and raised the possibility of giving a talk at the university. I expressed skepticism that the talk would have any effect, but wrote I was afraid that "ten or twenty years from now I may look back and say that I saw and understood what was going on, but I didn't try to speak up."

Van Allen was ninety years old, and I had qualms about putting him on the spot. So the next morning, before sending the letter, I called Don Gurnett to ask his opinion. Don had been Van Allen's best student in the 1960s and was now a physics professor at the university. He agreed that my proposed talk was appropriate and said he would discuss the matter with Van Allen.

My letter made clear the political sensitivity of the proposed lecture—a red flag to a university supported by public funds. The university president asked me, via Van Allen, to verify that I would be speaking as a private citizen and that government funds would not be supporting the trip. I sensed that it was only because of Van

Allen's legendary status in Iowa that my talk was approved as a university "Distinguished Lecture."

I did not expect my talk to alter votes in the upcoming election. Yet in the back of my mind I wondered: What if this public lecture leads to publicity and debate, and Professor Van Allen indicates agreement with my position? Given his reputation, it might influence fence-sitters in Iowa. At one time there had been a brouhaha about the safety of microwave ovens (manufactured in nearby Amana, Iowa). Van Allen denigrated that concern, offering to sit on a microwave oven while it cooked his dinner. That settled the matter for many people.

Initial plans had me appearing on the statewide *Talk of Iowa* radio program—a program that could have had more political effect than the lecture. Iowa is a "purple" state, sometimes voting Republican, sometimes Democratic. It was conceivable that Iowa might be pivotal in the presidential election. I had decided to mention my preference for John Kerry over George W. Bush, based on their positions about climate and energy. My on-air endorsement would have been lukewarm, though, as I had already indicated that I would have voted for John McCain if he had been on the ballot. (My enthusiasm for McCain, based on his crusade for campaign finance reform, dissipated when he backed away from that issue. The role of money in our capitals is the biggest problem for democracy and for the planet, in my opinion. For that and many other reasons, I voted for Barack Obama, not McCain, in 2008.)

However, the invitation for the *Talk of Iowa* interview was withdrawn, ostensibly because there had not been enough time to arrange it. I presumed that it was really because the university did not want to get embroiled in politics. But I was not about to complain. Professor Van Allen had done everything I possibly could have expected by getting approval for a "Distinguished Lecture" with little lead time.

I decided to write out my presentation and read it aloud before the audience—not an ideal format, but it would be worse to fumble around and forget important points. Besides, I could send the written version to a few people in the media. Because I was speaking in Iowa City rather than Washington, a printed copy provided the best hope for reaching a broad public.

I sent a draft of my presentation to Andrew Revkin of the *New York Times* on October 25, the day before my speech. Revkin wrote an article titled "NASA Expert Criticizes Bush on Global Warming Policy" that appeared in the October 26 issue of the newspaper. By then Anniek and I were on a plane to Iowa.

A call from my colleague Larry Travis greeted us at our hotel, relaying phoned and e-mailed warnings from Andrew Falcon of the NASA headquarters Chief Counsel's Office. Falcon's e-mail read:

> If the Times article is correct, he [Hansen] is going further than he has in the past, thereby placing himself at significant personal risk this evening. NASA does not have any control, input, or even insight into a decision by the Office of Special Counsel to prosecute Hatch Act violations. [The Hatch Act is the federal law that restricts participation by government employees in many aspects of partisan political activity. It does not limit the right to vote and express opinions about candidates and issues.] . . . In view of the timing, the content of the Times article, and the current flap with the Office of the Special Counsel [about a possible Hatch Act violation at Kennedy Space Center], I think it would be advisable for Dr. Hansen, for his own sake, to consider modifying his speech to eliminate the political elements. This is not a direction from NASA management, which has not expressed an opinion on the speech, at least not to my knowledge. This is simply offered in my capacity as an ethics counselor solely for the purpose of ensuring that, whatever Dr. Hansen does, he understands the ramifications of it, and, if he deems it appropriate, takes steps to mitigate the risk to himself.

That advice did not seem credible. I was using vacation time for the trip and was paying for the hotel and airfares. I was not about to change my mind. I had already gone through a struggle with self-doubt that chilly morning at 4:30 A.M. as I was dodging traffic while crossing the street on my way home after working extra late, realizing that I would get only an hour's sleep. Why, I wondered, was I doing such an almost surely fruitless thing? Wasn't I tilting at windmills?

It would be nice, for the sake of this book, if I had thought of my grandchildren at that moment. Instead, I thought of a cryptic four-word enigma that had stuck with me for decades. It was advice from Donald Hunten, who, along with Richard Goody, had been the father of the Pioneer mission to Venus. Hunten is small in stature but very authoritative. He speaks with a gravelly voice, seeming to push the words out from deep within his throat. I presumed Hunten had been responsible, at least in part, for the selection of our experiment to measure the Venus clouds as part of the Pioneer mission. Thus in 1978, when I wanted to resign as a principal investigator on Pioneer Venus so I could study Earth's climate full-time, I felt that I should seek Hunten's approval. I remember his advice as four gruff words: "Be true to yourself." What did that mean? Venus or Earth? I was not about to query him further.

I told that story later in a commencement address at my hometown high school in Denison, Iowa, suggesting that the students make their own interpretation. Perhaps Hunten only wanted me to think, to be sure that what I did was consistent with values I would like to have.

In any case, I had written my talk to be given at the University of Iowa carefully and was not going to change it. My aim was to

FIGURE 8. *Faustus contemplates benefits of a bargain with Mephistopheles. Humans made their own Faustian bargain via fossil fuel addiction. Time for possible redemption runs short.*

explain the science as well as I could, and also make clear the way things were working in Washington.

My Iowa talk was titled "Dangerous Anthropogenic Interference: A Discussion of Humanity's Faustian Climate Bargain and the Payments Coming Due." Humanity's Faustian bargain with fossil fuels, I suggested, has more far-reaching consequences, and, as we'll see in later chapters, some protagonists play a more shameful role than was the case in the bargain the grasping Dr. Faustus struck with the devil.

My talk included a discussion of the way science works, information on the greenhouse effect from Venus and Mars, an evaluation of climate sensitivity from paleoclimate data, and an analysis of ongoing climate change. The full version of the talk is on my Web site. Here I include observational data updated to 2009, and I focus on the most important uncertain aspect of the global warming story: the degree to which aerosol cooling is offsetting greenhouse gas warming—the Faustian climate bargain that humanity inadvertently entered into via the aerosol cooling effect.

The Faustian aerosol bargain arises from the simultaneous production of greenhouse gas warming and aerosol cooling, both primarily a result of fossil fuel burning. In a 1990 article in *Nature*, Andy Lacis and I described this, pointing out that aerosol cooling can continue to offset a large fraction of greenhouse warming only if particulate air pollution continues to increase rapidly. But at some point fossil fuels will run out, or people will get fed up with increasing air pollution and decide to clean up particulate pollution. Then, because greenhouse gases remain in the air for centuries and aerosols

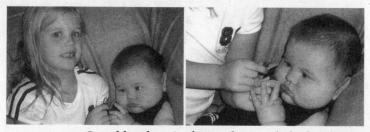

FIGURE 9. *Granddaughter Sophie explains to baby brother Connor that the net climate forcing is equivalent to having two 1-watt lightbulbs over each square meter of Earth's surface. Connor, however, counts only 1 watt.*

fall out within days after aerosol emission stops, the payment—via rapid increase of global warming—will come due.

I use photos of my first two grandchildren (figure 9) for quantitative discussion of the Faustian aerosol bargain. Sophie is explaining to her younger brother Connor that the net climate forcing is about 2 watts—she holds two miniature 1-watt lightbulbs. Two watts is approximately the sum of the change of all estimated climate forcings between preindustrial time and the first decade of this century (see figure 1, page 6). Connor, however, seems to count only 1 watt.

Connor could be right. The problem is that we do not have measurements for the climate forcing caused by human-made aerosols, that is, fine particles in the air. Greenhouse gases, in contrast, are measured very precisely. As a result, we know that the greenhouse gas forcing is close to 3 watts, as shown in figure 10. But the aerosol forcing could be anywhere in the range of –3 watts to near zero, as represented by the dashed curve, which is a crude estimate for the probability of a given aerosol forcing.

The net human-made climate forcing—warming due to greenhouse gases, offset by uncertain aerosol cooling—is represented by the solid area in figure 10, which is the probability distribution for the net forcing. The most likely net forcing is close to Sophie's 2 watts. But there is a substantial chance that Connor's 1 watt could be closer to the truth. Does it matter which one is right, or closer to being right? Yes, it matters a lot!

If the net forcing is 2 watts, aerosols have been masking about one third of the greenhouse forcing. So if humanity makes a big effort to clean up particulate air pollution (say, reducing human-made aerosols by half), the net forcing will increase by only a quarter, from 2 to 2.5 watts. The additional global warming would not be welcome, but it might not be earthshaking.

On the other hand, if the net forcing is only 1 watt, that is, if aerosol forcing is –2 watts, that means aerosols have been masking most of the greenhouse warming. In that case, if humanity reduces particulate pollution by even half, the net climate forcing would double. That increased forcing, combined with a continued greenhouse gas increase, might push the planet beyond tipping points with disastrous consequences. The current smaller net climate forcing already is causing a notable recession of mountain glaciers around the world, affecting freshwater availability, shifting climatic zones,

Greenhouse Gas, Aerosol & Net Climate Forcing

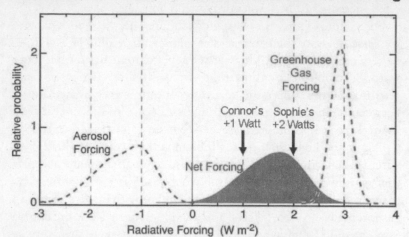

FIGURE 10. *Climate forcings by human-made greenhouse gases,*
aerosols, and their net effect. The greenhouse gas forcing is
3 watts (per square meter) with only small uncertainty, but the aerosol
forcing is very uncertain, as represented by the broad probability
function. Thus either Sophie's 2 watts or Connor's 1 watt is within
the range of likely net forcing. (Adapted from IPCC, Climate Change
2007: The Physical Science Basis. *See sources for chapter 1.)*

increasing fires and flooding, promoting the loss of Arctic sea ice and
vulnerable coral reefs, accelerating mass loss from the Greenland
and Antarctic ice sheets with rising sea level, and putting pressures
on many species, leading to a danger of mass extinctions.

The importance of knowing the actual aerosol forcing is thus obvious. The missing aerosol measurement was the principal objective of the Climsat mission, which I proposed at the Gore-Mikulski
roundtable meetings in 1989 and 1990. Measurement of the aerosol
forcing involves determining the effect of aerosols on clouds via
precise simultaneous polarimetry data for reflected sunlight and in-
terferometric data for emitted thermal radiation.

The satellite mission never took place, but little is gained by
crying over spilled milk. What we can do, in the absence of adequate aerosol measurements, is look instead for a measurable signature of the net climate forcing. The most fundamental effect of
net climate forcing is on Earth's energy balance. If the net climate
forcing is positive, Earth must be gaining more energy (as absorbed

sunlight) than it is losing (as emitted heat radiation). If the climate forcing stabilizes, the energy imbalance gradually declines as the planet warms up in response to the forcing and increases its heat radiation to space. The energy imbalance remaining at any time reveals the portion of the net climate forcing that has not yet been responded to.

So if we measure Earth's current energy imbalance, we can determine the amount of global warming still "in the pipeline." Direct measurement of the imbalance would require continuous monitoring by several satellites measuring radiation outgoing in all directions to an absolute accuracy of about a tenth of a watt. That is impractical. But a precise measurement can be inferred indirectly— from the rate at which heat is being stored in available reservoirs on Earth. The dominant reservoir, by far, is the ocean. For this reason I have argued for the past two decades that the single most important geophysical measurement is change of ocean heat content. If we can measure how much the oceans are warming, we will know not only how much additional global warming is in the pipeline but also how much we must reduce the human-made climate forcing if we want to stabilize climate.

In 1997 a number of colleagues and I published a paper ("Forcings and Chaos in Interannual to Decadal Climate Change") that concluded Earth was out of energy balance by at least +0.5 watt. Our conclusion was based on a comparison of climate model simulations with analysis of global ocean temperatures by the National Oceanic and Atmospheric Administration's Sydney Levitus. The result was tentative because ocean temperature measurements were spotty, especially in the deep ocean and at high latitudes, and because it was uncertain whether the temperature data were sufficiently accurate. Large, probably unrealistic temporal fluctuations in Levitus's analyzed ocean heat content data suggested the possibility of errors due to instrumental changes or incomplete sampling of the ocean.

In my Iowa talk I showed a new analysis of ocean heat, by Josh Willis and colleagues, that indicated heat storage at a rate of 0.6 watt in the upper 750 meters of the ocean during 1993–2003. While that result agreed well with the energy imbalance in our climate model simulations, Catia Domingues and colleagues later showed that instrumental biases affected the results of both Levitus's and Willis's analyses. Until instrumental issues are resolved and good heat storage data is obtained for the entire ocean, it is not possible to infer the net climate forcing acting on Earth.

The difficulty of determining the precise value of Earth's energy imbalance based on existing measurements does not diminish its importance. Earth's energy imbalance is our best indication of where global temperature is headed and of how much global forcings must be altered to stabilize climate. We know for sure that the ocean is warming and that the planet is out of energy balance, averaged over the past decade or two (averaging minimizes the effect of cyclic solar variability and chaotic atmospheric and ocean variability). But, given the imprecision of the measurements, the imbalance is only known to be somewhere in the range of 0.25 to 0.75 watt. This includes about 0.1 watt for changes in heat reservoirs other than the ocean, i.e., heat that goes into warming the air, lakes, and continents and melting sea ice, land ice, and ice shelves.

So the actual planetary energy imbalance is probably less than the energy imbalance that was calculated in the climate simulations we made in 2004 and 2005—which were the climate model results we submitted to the IPCC assessment that was published in 2007. Those climate model results showed the planetary energy imbalance in the first several years of the twenty-first century, averaged over a series of model runs, to be 0.75 watt. What can we learn from the discrepancy, assuming it is not due to observational error or just the variability of the climate system? We know the discrepancy cannot be due to an error in our model's climate sensitivity, since the paleoclimate data (discussed in chapter 3) confirm the climate sensitivity of 3 degrees Celsius for doubled carbon dioxide. Instead, the discrepancy has implications about the assumed net climate forcing.

Our climate simulations assumed that the amount of atmospheric aerosol has stayed constant since 1990, with the hope that aerosol reductions in the United States and Europe, due to clean air regulations, would tend to offset aerosol increases in developing countries such as China and India. However, clean air rules in the West, and reduced Russian emissions due to economic collapse, occurred largely prior to 1993. Air pollution in developing countries, on the other hand, surely increased during the past two decades, a conclusion supported by global measurements of decreased atmospheric visibility. So an increase of aerosols in the period from 1990 to 2009 is one candidate for explaining the observed planetary energy imbalance.

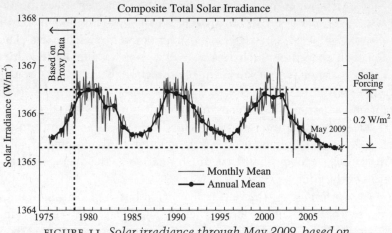

FIGURE 11. *Solar irradiance through May 2009, based on concatenation of multiple satellite records by Claus Fröhlich and Judith Lean. (Data from Fröhlich, "Solar Irradiance Variability Since 1978." See sources for chapter 1.)*

Greenhouse gases and aerosols are the two largest human-made climate forcings. But there is a third significant climate forcing we need to look at when we use the planet's energy imbalance to check on the status of the net climate forcing, and when we try to assess likely climate change in the next few years: solar variability. So let's consider the sun's role in climate change.

Indeed, there are many people, including scientists, who believe that the sun is the most important factor in climate change, the dominant climate forcing. It is easy to understand their suspicions. Earth gets its warmth from the sun. The sun is variable. Correlations of solar variability and climate change are well known. But what we need is an objective, quantitative comparison of solar and other climate forcings.

Precise monitoring of solar irradiance, the amount of solar radiation reaching Earth, began in the late 1970s. The data, shown in figure 11, reveal that the sun was dimmer in 2009 than at any other time in the period of accurate data. Moreover, the current solar minimum has lasted longer than earlier minima in the satellite era. This solar variability affects Earth's energy balance and global temperature. It also can provide a climate test that can help us refine our understanding of climate change in coming years and decades.

Solar irradiance varies by about a tenth of a percent over the average ten-to-twelve-year solar cycle, as shown by figure 11. Earth absorbs 240 watts of sunlight per square meter of its surface. (The absorbed energy is much less than the sun's irradiance that hits Earth perpendicular to the Earth-sun direction, because the circular cross section of Earth presented to the sun is a quarter of Earth's surface area, and 30 percent of incident sunlight is reflected to space without being absorbed.) Thus the sun at solar minimum causes a forcing of about –0.2 watt, relative to the forcing at solar maximum, or about half that much relative to the average brightness of the sun.

A solar forcing of –0.2 watt is significant but not a dominant forcing. By contrast, the carbon dioxide forcing today, relative to preindustrial times, is about 1.5 watts. However, much of the carbon dioxide forcing has already been "used up" in causing the warming of the past century. It is more relevant, then, to compare the solar forcing with the planet's energy imbalance, which we estimated as about 0.5 watt averaged over the past two decades. But before judging the sun's importance, we should note that there are mechanisms by which the sun's effect may be magnified or diminished.

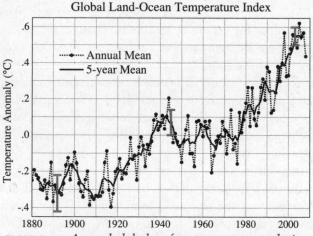

FIGURE 12. *Annual global surface temperature relative to 1951–1980 mean. Vertical bars at several points show estimated 95 percent confidence range. (Updates of data from Hansen et al., "GISS Analysis of Surface Temperature Change." See sources.)*

One of the papers I was working on in 2004, "Efficacy of Climate Forcings," addressed this matter. In that paper, we showed that it is not sufficient to know only the magnitude of a forcing in watts, because some forcing mechanisms have a greater "efficacy" than others. For example, we showed that the effect of solar irradiance forcing is reduced about 10 percent, relative to the standard carbon dioxide forcing, because solar forcing is greatest at low latitudes, where there is little amplifying feedback from ice or snow. But solar forcing is increased about 20 percent by an indirect effect—the large solar variability at ultraviolet wavelengths alters atmospheric ozone. These known effects yield a net efficacy of about 110 percent for solar forcing relative to an equal carbon dioxide forcing—an amplification too small to substantially alter our assessment of the sun's role in climate change.

So, if solar variability is to be a more significant climate forcing, there must be another, larger, indirect effect of the sun. The favorite among solar aficionados is an almost Rube Goldberg effect of galactic cosmic rays (GCRs) that goes like this: At solar minimum, GCRs penetrate farther into Earth's atmosphere and increase atmospheric ionization, the ions serve as condensation nuclei for clouds, cloud cover increases, and the clouds reflect sunlight and cause global cooling. Indeed, it is true that the solar cycle has an effect on atmospheric ionization, but the state of the science does not yet allow definitive quantitative evaluation.

Fortunately, we have a way to skirt such difficult theoretical problems. We can go straight to empirical data to evaluate the effect of solar climate forcing, even its indirect effects. All we need are the measurements of solar variability from recent decades and observed global temperature.

The observed temperature curve (figure 12) does not overtly display solar cycle variability. However, statistical analysis reveals a clear correlation. Ka Kit Tung and Charles Camp carried out the most sophisticated and accurate comparison of solar irradiance and global temperature, finding a global warming of 0.16 degree Celsius for a solar irradiance change of 0.2 watt. That corresponds to 0.8 degree Celsius for each watt of forcing, or 3.2 degrees Celsius for a doubled CO_2 forcing of 4 watts. At first glance, Tung and Camp's climate sensitivity results approximately match the climate sensitivity of 0.75 degree Celsius per watt, or 3 degrees Celsius for doubled CO_2,

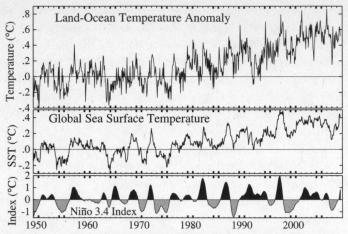

FIGURE 13. *Monthly global (land plus ocean) and global ocean surface temperature relative to the 1951–1980 mean. The land-plus-ocean graph is noisy because of weather variability. The bottom diagram is the Niño 3.4 index for tropical Pacific Ocean temperature. (Top figure data from Hansen et al., "GISS Analysis of Surface Temperature Change." See sources for chapter 6. Middle figure data from Hansen et al., "Target Atmospheric CO₂." See sources for chapter 8. Bottom figure data from NOAA Climate Prediction Center.)*

that we derived from paleoclimate records. However, the paleoclimate result refers to the long-term climate response, that is, after the ocean has had enough time to adjust to the changed climate forcing. During the ten-to-twelve-year solar cycle, surface air temperature would only achieve 50 percent of its equilibrium response, with an uncertainty of about 10 percent. Thus Tung and Camp's results imply that there is an indirect solar forcing that doubles the climate impact of the direct solar forcing—an efficacy of 200 percent. So maybe the solar aficionados are at least partly right—the sun's effect is larger than implied by the irradiance change alone, with magnification due to either galactic cosmic rays or some other unknown mechanism.

But before leaping to the conclusion that there are mechanisms amplifying the solar forcing by a factor of two, we must look a bit more closely at Tung and Camp's analysis. First, their method of avoiding volcanic influence on the record, removing two years of temperature data following two of the three large eruptions, neglects the residual longer-term effect of large volcanoes—coincidentally, the three large volcanoes in the period of study, Agung, El Chichón, and Pinatubo, all served to enhance cooling during solar minima. Second, Tung and Camp did not use the observed global temperature record of figure 12; instead they used a temperature record generated by a computer model in a process termed "reanalysis." Reanalysis uses a global atmospheric simulation of the past few decades constrained by inserting available observations, such as tropospheric temperatures measured by satellite. The global surface temperature resulting from the reanalysis is qualitatively similar to that shown in figure 12, but the solar signal is about one-third stronger when measured this way. Tung and Camp suggest that variations of global temperature used in conventional analyses, as in figure 12, which are based on observations at meteorological stations, are more muted than the variations resulting from their reanalysis method, because of the absence of observing stations in places such as the Arctic and the Sahara, the areas with the largest temperature variability. Their supposition is plausible, but it is possible that instead their reanalysis model magnifies the temperature change.

The bottom line is that the only thing we can say with confidence is that the effective climate forcing due to the ten-to-twelve-year solar cycle has an amplitude in the range of 0.2 to 0.4 watt. In wonkish terminology, the 0.2 watt solar forcing has an "efficacy" between 100 and 200 percent.

Now that we have those figures, we can compare the natural solar climate forcing and the human-made carbon dioxide forcing. The annual increase of carbon dioxide today is about 2 ppm, which causes an annual increase of about 0.03 watt, with efficacy identically 100 percent. Thus even if the efficacy of the solar forcing is 200 percent, and if the sun's brightness remains at the 2009 solar minimum value for a long period, the cooling effect relative to the average solar irradiance would be offset in seven years by a continuing carbon dioxide increase at recent rates. So there is no chance whatsoever that the sun can cause Earth to go into a new Little Ice Age—the numbers above confirm that human-made forcing now overwhelms the natural climate forcing.

Why, then, am I bombarded by demands to repent, to admit that

global warming is a hoax? I have received scores of messages claim-
ing that humans are not responsible for climate change and that
Earth is headed into much colder conditions. Usually it is claimed
that the sun controls climate and that the sun is moving into a period of reduced luminosity. As "proof," these messages are often
accompanied by a graph showing recent global temperature and
an assertion that already "half of the global warming of the past
century" has been lost. There is remarkable similarity among the
messages, and most end by demanding that I resign from the government.

These people seem to be emboldened by the fact that 2008 was
a cool year, the coolest year since 2000. Some insight into the
2008 cooling is provided by a comparison of monthly mean global
(land plus ocean) surface temperature, global ocean temperature,
and the oceanic Niño index (figure 13). The index is positive during El Niños, when the equatorial Pacific Ocean is warm, and
negative during the cool La Niña phase. This natural dynamical
oscillation of Pacific Ocean temperatures has a big impact on global
temperatures. The coolness of 2008 is associated with a strong La
Niña.

As the Pacific Ocean has moved into the El Niño phase in 2009,
I expect global temperature to move back to record or near-record
levels. There is a lag of a few months between the Niño index and
global temperature, so 2010 should be a year that is back to near-
record global temperature levels, dispelling the notion of a coming
ice age. We are still at solar minimum in the sun cycle, though,
which does drag down the temperature, so the next El Niño may
produce merely a near-record, not a record, global temperature.

A more important measurement during the next several years
will be ocean heat content. There is an improving global network
of ocean floats with regular yo-yo temperature probing of the upper
750 meters of ocean, along with limited measurements at greater
depths. These should provide a better measure of ocean heat uptake
and thus the planet's energy imbalance. If solar irradiance begins to
pick up, as most solar physicists predict, I expect ocean heat uptake
to reflect that change. Measurements of heat storage over the next
solar cycle, especially if global aerosol measurements are also obtained, have the potential to help confirm our understanding of the
state of the planet, the causes of climate change, and the change of

global climate forcing that is needed to restore the planet's energy
balance and stabilize climate.

BACK TO IOWA CITY. My talk, on the evening of October 26 in the
Van Allen Hall auditorium, seemed anticlimactic. I was pretty exhausted, after getting only an hour of sleep the night before, when I
read the talk, which lasted a good hour. In a photo taken by the university, which Mark Bowen used on the cover of *Censoring Science*,
it looked like I must have been trying to smile. The auditorium was
nearly full—the audience was sympathetic, and there were not
many questions. James Van Allen sat in the front, his wife, Abigail,
in the back. Afterward he invited me to come to his office the next
morning.

When I met him there, he gave me the verdict on the talk from
the judge: Abigail said that it was good and understandable, but she
thought that it would be very difficult to get the public to pay any
attention, given issues with health care and many other matters
that would have higher priority. Van asked if the talk had achieved
what I intended—which of course I affirmed.

As I sat in his office, I wondered how in the world I could ever
have been intimidated by such a kind and gentle man. He gave me
a copy of an op-ed piece that he had recently written, making the
case for robotic space exploration and termination of the manned
space program. I had never agreed with his position on that subject,
but there was no reason to debate it. He asked if there was any
progress on the aerosol measurements I had proposed fifteen years
earlier, the polarimeter and interferometer. I could report that work
had started on the polarimeter, but I was skeptical about whether it
would ever be completed, given likely reactions to my current criticisms. It was the last time that I saw Van.

Articles describing my talk appeared in a few newspapers. CNN
.com quoted me as saying, "In my more than three decades in government, I have never seen anything approaching the degree to
which information flow from scientists to the public has been
screened and controlled as it is now." The Associated Press reported
that I had said the administration wanted to hear only scientific results that "fit predetermined, inflexible positions," which I described as "a recipe for environmental disaster."

There was still a week before the election. I was given a last hope for bringing public attention to the matter when I received a call from a reporter at National Public Radio (NPR). The reporter had obtained a copy of a memo to NASA employees from Glenn Mahone, head of NASA's Office of Public Affairs. He wondered whether there was something nefarious about the memo, as it concerned procedures aimed at maintaining "consistency." I concurred that the intent seemed to be to keep everybody on a predetermined message, a dangerous approach for a science agency—indeed, it seemed reminiscent of the Catholic Church and Galileo.

I told him that NASA press releases were being funneled to the White House for approval or changes. I suggested that it would be very useful if NPR could make clear what was happening and that I might be able to provide contacts who would confirm the story. Specifically, I had spoken with Rob Gutro and his colleague Krishna Ramanujan on the day before my Iowa talk. They seemed scared stiff, which was understandable, as they had young families to support. They said that I was "not going to get cooperation" from them, they had "been talked to," and they "could be fired." They explained that Glenn Mahone had driven from NASA headquarters to the Goddard Space Flight Center and had personally chewed them out in front of their superiors. Gutro and Ramanujan were not even NASA employees; they were contractors working for the Goddard Public Affairs Office. Mahone's action, to say the least, seemed highly inappropriate.

I urged NPR to pursue the matter, said that I would cooperate, and provided contact information for both Gutro and Ramanujan. Despite their reservations about possible retribution, I thought that they might be willing to cooperate if NPR provided assurance that the full story would come out. I never heard anything back from NPR.

On Election Day, I made the short drive down the road to the polling place, at a junior high school. I live in Pennsylvania, another purple state, so my vote would count. That evening Anniek and I made popcorn and settled in to watch the returns. Iowa was interesting—it was too close to call until the next day, decided by a handful of votes (Bush won). But the election hinged on Ohio, and by midnight it was becoming more and more clear that Bush had won Ohio and would be president for four more years.

I wanted to be in my office the next day in case anybody had concerns over the election's implications—given my public endorsement of Kerry. Also I wanted to send a memo to the president's science adviser. So as Bush's victory became nearly certain, we decided to return to New York City late that night.

It is a half-hour drive over two-way roads between our house and the interstate highway. As we came around a curve, suddenly there was a deer in front of us. I hit the brakes, losing steering control, unable to react fast enough. We slammed into the deer, whose body was hurtled down the road. We sat stunned for several seconds. The deer lay motionless, apparently dead. Then, at age sixty-three, for the first time since childhood, I burst into tears. I am not sure if I was crying for the deer, the nation, or the planet.

CHAPTER 7

Is There Still Time? A Tribute to Charles David Keeling

IN ORDER FOR A DEMOCRACY TO FUNCTION well, the public needs to be honestly informed. But the undue influence of special interests and government greenwash pose formidable barriers to a well-informed general public. Without a well-informed public, humanity itself and all species on the planet are threatened. That is a strong assertion, but I hope the remaining chapters of this book leave you convinced of its validity.

The morning after the election I sent a letter to the White House Office of Science and Technology Policy, requesting an opportunity to discuss the state of global warming science with the president's science adviser, John Marburger. I received no response, which was not surprising, given my public criticism of the Bush administration.

I thought a public talk might be an alternative. My talk in Iowa City had been before a single university audience, and maybe in a different setting I could do a better job of making the science story clear. An opportunity for a public statement arose in August 2005, when Ralph Keeling asked me to give a lecture in honor of his father, Charles David Keeling, at the American Geophysical Union meeting in San Francisco in December. David Keeling, who died that June, is famous for the painstaking, precise observations of atmospheric carbon dioxide that he initiated in 1957. He doggedly continued this monitoring for decades, continually fighting bureaucratic obstacles. In so doing, David Keeling brought to the world's attention the reality of rising atmospheric carbon dioxide levels.

Giving a talk honoring David Keeling would be a privilege, but I questioned whether I was the best person to address geochemistry and the carbon cycle. Ralph responded that he had read a copy of my Iowa talk and wanted to give me "the stage for presenting [my] perspective on the overall science of global warming and where we are heading." He said, "You'd honor my father best by telling your own story and thereby carrying forward the torch that he helped to light." Ralph noted that only minutes before dying of a heart attack, David Keeling was involved in a discussion with one of his other sons about my paper "Earth's Energy Imbalance," which had just appeared in the journal *Science*.

The timing seemed right. I had been working for more than a year on the paper "Is There Still Time to Avoid 'Dangerous Anthropogenic Interference' with Global Climate?" A one-hour lecture at the American Geophysical Union meeting, the largest conference in geosciences, would deserve attention if it were backed by a solid scientific paper. As it turned out, for somewhat different reasons, even the White House took notice.

I worked hard on the paper for months. It was nearly in final form a week before the meeting, when I had to set it aside and start preparing my talk. I was still working on the talk the day before the meeting, sitting on the floor at JFK airport, my laptop computer plugged into the nearest electrical outlet—with Anniek at the gate a few hundred yards away, keeping tabs on the status of our delayed flight.

I was struggling with the bottom line, the summary: Should a scientist connect the dots in the climate story all the way to policy implications? My experiences in the five years since we published our alternative scenario paper provided some relevant perspective. In that paper we showed that if global fossil fuel emissions peaked early in the twenty-first century and then declined steadily, the amount of carbon dioxide in the atmosphere could be kept below 450 parts per million and climate change might be tolerable. We argued that such a path was technically feasible, because of the great potential of energy efficiency and non-carbon energy sources, but their ascendancy would require appropriate policies, especially an increasing price on carbon emissions.

Based on my encounters with the vice president's Task Force and the Council on Environmental Quality, the trials of the automobile manufacturers versus California and Vermont, and meetings such as the one I attended at ExxonMobil headquarters, I had an empirical basis for inferences about obstacles to needed policies. Thus, I tentatively wrote my concluding paragraphs:

> If an alternative scenario is practical, has multiple benefits, and makes good common sense, why are we not doing it?
>
> There is little merit in casting blame for inaction, unless it helps point toward a solution. It seems to me that special interests have been a roadblock wielding undue influence over policymakers. The special interests seek to maintain short-term profits with little regard to either the long-term impact on the planet that will be inherited by our children and grandchildren or the long-term economic well-being of our country.
>
> The public, if well informed, has the ability to override the influence of special interests, and the public has shown that they feel a stewardship toward Earth and all of its inhabitants. Scientists can play a useful role if they help communicate the climate change story to the public in a credible, understandable fashion.

I was hesitant because I could be getting into a quagmire. Scientists, politicians, and the special interests all would advise me to stick strictly to the science. But scientists are equipped to connect all the dots in an objective way. Do they have an obligation to do that? Can scientists maintain their scientific objectivity if they get involved in policy-related discussions? And after this one additional talk, could I get back to just doing science?

Coincidentally, I had met a good sounding board for such questions a few months earlier: Bill Blakemore of ABC Television. We had participated in a discussion at Wesleyan University, in which we agreed that the media needed to do a better job of informing the public about human-made climate change. Bill had arranged for me to give a presentation to the president of ABC News and other senior staff members in mid-November about the current understanding of climate change. During the discussion at ABC I promised to provide them my "Keeling" talk when it was ready.

Blakemore is a gentle, expansive person who feels like a close friend after one conversation. His broad experience and curiosity also imply a certain wisdom, which is probably why I included an implicit question when I e-mailed him my draft presentation before my American Geophysical Union talk in San Francisco: "I do not intend to make it political, but I think I need a couple of sentences regarding special interests, to help explain why we are not taking sensible steps. I'm still struggling with this aspect, how to remain an objective scientist."

Blakemore responded that it struck him that the last two paragraphs were "statements of fact or of your belief and do nothing to detract from your function as an objective scientist. This is, after all, a story and a scientific object of study about what human action has done and is doing to the planet, so your statements in these last two paragraphs are not even out of place regarding the scientific subject you present. All communication is biased. What makes the difference between a propagandist on one side and a professional journalist or scientist on the other is not that the journalist or scientist 'set their biases aside' but that they are open about them and constantly putting them to the test, ready to change them. I base this on a close study of the modern philosophy of science, especially that of Karl Popper and Peter Brian Medawar and this is all aside from the fact that no one is going to object to a serious scientist trying to alert the public about the import of alarming news."

There is a counterargument against my explicit criticism of "special interests." Mahatma Gandhi warned his followers to be "most careful about accusing the opponent of wickedness . . . Those who we regard as wicked as a rule return the compliment." And I realize that the captains of industry must be a big part of the global warming solution; the needed changes of energy infrastructure require their leadership. But it is hard to find solutions if we do not paint the picture accurately. So I decided to retain the concluding paragraphs.

The heart of my "Keeling" presentation described "multiple lines of evidence indicating that the Earth's climate is nearing, but has not passed, a tipping point, beyond which it will be impossible to avoid climate change with far-ranging undesirable consequences." I concluded that it was necessary to begin fundamental changes to the energy infrastructure within a decade, so that global carbon dioxide emissions would stabilize in the first quarter of the twenty-first century and decline in the second quarter. This would allow the

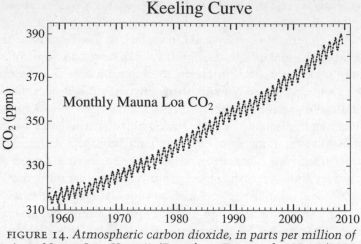

FIGURE 14. *Atmospheric carbon dioxide, in parts per million of air, at Mauna Loa, Hawaii. (Data from Tans et al., NOAA/ESRL Web site, http://www.esrl.noaa.gov/gmd/ccgg/trends/.)*

atmospheric carbon dioxide amount to peak in the neighborhood of 450 or 475 ppm. If other greenhouse gases were reduced by feasible amounts, additional global warming could be kept at less than 1 degree Celsius. I argued that this proposed scenario was technically feasible, but it would require strong policy leadership and international cooperation. I pointed out the multiple benefits of such an energy strategy, for human health, clean air and water, and national security.

It is especially appropriate that this talk was in honor of David Keeling. Keeling developed a technique that greatly reduced the average error in measuring carbon dioxide in the atmosphere. He realized that the changes he was observing in the amount of carbon dioxide were systematic and real, not due to measurement error. Keeling then set upon making measurements with the focus and dedication of a scientist who knows that he is onto an important problem and has the tools to address it. Despite being a loving family man, he missed the birth of his first child because he religiously went out every four hours to measure carbon dioxide.

Keeling's measurements near his home or laboratory in California showed that the amount of carbon dioxide in the air decreased during the day as trees and other vegetation assimilated carbon dioxide during the process of photosynthesis. At night, atmospheric carbon dioxide increased as plants respired. Human-made sources of carbon dioxide, mainly fossil fuel burning, also affected observations, depending on the proximity of sources.

Keeling needed to sample air in remote locations to investigate global carbon dioxide changes. So he set up acquisition of daily air samples at Mauna Loa in Hawaii, and soon thereafter at the South Pole. The air Keeling was sampling on Mauna Loa was pristine Pacific Ocean air, high in the atmosphere, brought to Hawaii by westerly winds, uncontaminated by local human sources.

The now-famous Keeling curve revealed annual oscillations and an average carbon dioxide amount that increased every year. The oscillations could be readily traced to the dominance of northern hemisphere vegetation, which draws down atmospheric carbon dioxide during the northern hemisphere growing season and replenishes it as plant litter decays during the autumn and winter. As summarized by Mark Bowen in *Thin Ice*: plants, in effect, take one breath a day, and Earth overall takes one breath a year.

Keeling's curve had a big impact because it confirmed that carbon dioxide was increasing year by year. This result was not a surprise, but the precision of the data brought increased attention to carbon dioxide and concerns about possible climate effects. Also, as Keeling's record grew longer, it became clear that the magnitude of the annual carbon dioxide growth was getting larger and larger. The Keeling curve in figure 14 may seem to be increasing almost along a straight line, but it is far from that. Indeed, the annual carbon dioxide increase is now about three times greater than it was when Keeling began his measurements in 1957.

The annual increase of global mean carbon dioxide is one of the key quantities that we must keep our eye on to understand the state of the climate and prospects for the planet's future. I discuss these key quantities in the afterword, and I will keep these quantities updated on the Web site identified there.

Interpretation of Keeling's measurements requires knowledge of the "carbon cycle," the movement of carbon among the atmosphere, biosphere, soil, and ocean reservoirs. The carbon cycle is summarized in figure 15. The unit of measure is a billion metric tons of carbon, also called a gigaton of carbon, abbreviated as GtC; 1 ppm of carbon dioxide in the atmosphere is about 2.12 GtC.

The atmosphere today contains about 800 GtC as carbon dioxide. Plants contain about 600 GtC, primarily the wood in trees. Soils

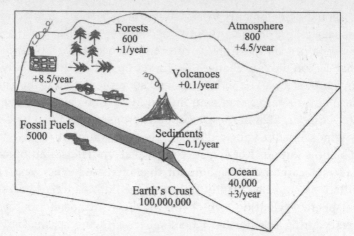

FIGURE 15. *Global carbon cycle (units are gigatons, each equal to a billion metric tons).*

contain about 1,500 GtC, which is mainly humus, decomposed organic matter. There is almost 40,000 GtC dissolved in the ocean.

Large, natural back-and-forth fluxes of carbon pass among these reservoirs. Plants take up carbon dioxide in photosynthesis, but plants and soils rapidly respire a similar amount of carbon dioxide back to the atmosphere. Carbon dioxide dissolves into cold ocean regions, but a similar amount is released to the atmosphere at other places. These uptakes and losses nearly balance over the year, but small imbalances occur and provide important climate feedbacks. For example, in interglacial-to-glacial climate change, as the ocean becomes colder, it dissolves more carbon dioxide, causing the atmosphere and plants to contain less carbon dioxide, which then drives further cooling. Conversely, when Earth's orbit or the tilt of the spin axis cause melting of snow and ice, this increases absorption of sunlight, and the warming ocean and soil release carbon dioxide and methane. This greenhouse gas amplifying feedback, as I showed earlier, accounts for nearly half the glacial-interglacial global temperature change.

Humans alter this natural carbon cycle in two major ways: by the burning of fossil fuels and deforestation. The rate at which fossil fuel carbon dioxide is injected into the global atmosphere is known with reasonably high accuracy, because oil, gas, and coal are well-tracked international commodities. The error in annual global fossil fuel use is probably less than 10 percent, even though some governments may not accurately report internal coal uses or sales.

The solid line in figure 16 is the global carbon dioxide emission from fossil fuel use. Emissions increased from less than 2 GtC in 1950 to more than 8 GtC per year in the last few years. The growth rate of emissions was about 4.5 percent per year from 1950 to 1973. It slowed to about 1.5 percent per year between 1973 and 2003, but between 2003 and 2008 it averaged about 3 percent per year as coal use increased rapidly, especially in China. China's annual emissions now exceed those of the United States. However, because of the long lifetime of atmospheric carbon dioxide, the United States is responsible for about three times more human-made carbon dioxide in the air today than is China.

Deforestation is the second important human-made source of atmospheric carbon dioxide. However, the magnitude of the annual deforestation rate is not known accurately. Valuable insight into carbon cycle uncertainties is provided by the simple ratio of the two quantities that are well known: the annual increase of atmospheric carbon dioxide divided by the fossil fuel emissions in the

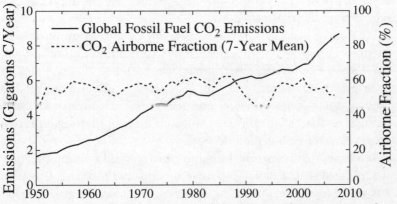

FIGURE 16. *Fossil fuel emissions and the fraction that appears in the atmosphere. (Emissions data from Boden et al., ORNL/CDIAC's Web site, http://cdiac.ornl.gov/trends/emis/meth_reg.html, and the fraction data are updates of Hansen and Sato, "Greenhouse Gas Growth Rates." See sources.)*

same year. This "airborne fraction" is the dashed curve in figure 16, shown in percent on the right-hand scale.

Remarkably, the airborne fraction, averaged over several years, has been nearly constant for fifty years at an average value of only 56 percent. In other words, even if we assume that there is no net deforestation, 44 percent of fossil fuel carbon dioxide is disappearing into sinks. Sinks are places—such as the ocean, forests, and soils—that can take up some of the excess carbon that humans are putting into the air. It is fortunate that sinks have been able to remove a significant fraction of the human emissions—otherwise the climate change would be larger. Recently, on the basis of both models and observations, the ocean is estimated to be taking up about 3 GtC per year. Thus, given the fossil fuel source of 8.5 GtC per year and the average atmospheric increase of 4.5 GtC, the total sink must be 4 GtC per year. Given the estimate of 3 GtC per year for the ocean sink, all other factors—mainly vegetation and soils—together must produce a net sink of about 1 GtC per year.

The fact that Earth's land masses continue to produce a net sink of carbon dioxide provides a glimmer of hope for the task of stabilizing climate. This carbon sink occurs despite large-scale deforestation in many parts of the world, as well as agricultural practices that tend to release soil carbon to the atmosphere. Improved agricultural and forestry practices could significantly increase the uptake of carbon dioxide, as we'll see later.

Any optimism, however, is dependent on the assumption that fossil fuel emissions will decline. If, instead, emissions continue to increase, the terrestrial system may become a less effective sink or even become a source of greenhouse gases. Some climate models predict, for example, that continued global warming will cause drought and forest fires in the Amazon, turning that region into a large source of carbon dioxide.

It follows that the world, humanity, has reached a fork in the road; we are faced with a choice of potential paths for the future. One path has global fossil fuel emissions declining at a pace, dictated by what the science is telling us, that defuses amplifying feedbacks and stabilizes climate. The other path is more or less business as usual, in which case amplifying feedbacks are expected to come into play and climate change will begin to spin out of our control.

Well, then, how can we evaluate as precisely as possible the path that humanity is beginning to travel at this critical time? One way is to keep our eyes on two key numbers with respect to the carbon cycle.

The first key number is the rate at which carbon dioxide is being pumped into the air by fossil fuel burning. This is shown by the solid line in figure 16.

A second number is needed to characterize the state of the carbon cycle, because the change of atmospheric carbon dioxide surely will not continue to average 56 percent in the long run; the percentage may either increase or decrease. The second key number defining the status of the carbon cycle is the annual growth of carbon dioxide in the air, shown in figure 17. This quantity reflects the combined effect of any change in the sinks for carbon dioxide and change of the net deforestation source of carbon dioxide.

This second key number is precisely known because of the monitoring that Keeling initiated. However, I must warn you that the value of this quantity fluctuates a lot from year to year, as shown in figure 17, so do not take observations in a single year as a basis for either alarm or rejoicing. One reason for the fluctuations is the oscillation of ocean surface temperature associated with the El

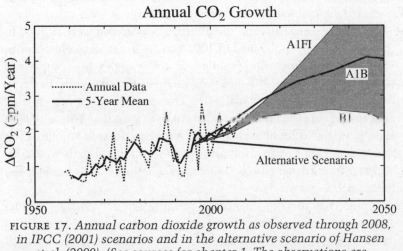

FIGURE 17. *Annual carbon dioxide growth as observed through 2008, in IPCC (2001) scenarios and in the alternative scenario of Hansen et al. (2000). (See sources for chapter 1. The observations are updates of Hansen and Sato, "Greenhouse Gas Growth Rates" (see sources), with original data from NOAA/ESRL Web site, http://www.esrl.noaa.gov/gmd/ccgg/trends/.)*

Niño–La Niña cycle, which affects the ocean's ability to absorb carbon dioxide. Also, droughts reduce the ability of vegetation to take up carbon dioxide, and forest fires release carbon dioxide.

The five-year mean is included in figure 17 to minimize the effect of fluctuations, but even this mean is quite variable. The largest fluctuation is the slow growth of atmospheric carbon dioxide in the early 1990s, which is probably an effect of the massive eruption of the Mount Pinatubo volcano in 1991. Pinatubo cooled the ocean for a few years, causing the ocean to dissolve more carbon dioxide. Volcanic aerosols also scatter incoming sunlight, making incoming light more diffuse (causing the sky to appear slightly milky in daytime). Plants grow better, and thus sequester more carbon, if the sunlight is diffuse rather than beating down on them from one direction.

As time goes on, it will become more and more useful to compare observed carbon dioxide growth with scenarios for the future, which are included in figure 17. All scenarios defined by IPCC have carbon dioxide growing faster and faster in the future. We shall see anon that all these scenarios yield climate disaster. Yet these scenarios are consistent with projections of government energy agencies, which universally assume, seemingly as a god-given fact, that all fossil fuels will be burned at a faster and faster rate. In contrast, our alternative scenario, also shown in figure 17, assumes that humanity is capable of exercising free will in determining its energy sources.

Both IPCC scenarios and the alternative scenario were defined in the late 1990s and published in 2001 and 2000, respectively. So now, real-world data from almost a decade are available for comparison with those scenarios. It is apparent that, so far, the world has continued on a business-as-usual path. Thus real-world carbon dioxide growth has exceeded that in our alternative scenario. Yet, as will be shown, it is still feasible to achieve a carbon dioxide amount even lower than that in the alternative scenario. Such a path would require restrictions on emissions from coal and unconventional fossil fuels, such as tar sands.

I hope that the key quantities defining climate change, and its causes and consequences, as summarized in the appendix and updated monthly on my Web site, will help the public understand climate change as it progresses and distinguish reality from propaganda and hyperbole. In that regard, the growth rate of atmospheric carbon dioxide in figure 17 provides a good example. It is apparent that, despite year-to-year fluctuations, carbon dioxide is growing at a rate in good agreement with the IPCC business-as-usual scenarios but faster than the alternative scenario.

This reality contrasts markedly with the impression created by the media. One frequently reads that greenhouse gases including carbon dioxide are increasing more rapidly than expected, emissions are exceeding expectations, ocean sinks have decreased, the soil has become a source of carbon dioxide, or deforestation has increased. A given story may have some basis in reality, but the net result is a misimpression, as figures 16 and 17 make clear.

Sometimes these stories appear simply because the media has a desire to find interesting "news" to report. Albert Einstein had misgivings about scientists describing research progress to the media, especially preliminary results, because inevitably, he said, this creates "the impression that every five minutes there is a revolution in Science, somewhat like a coup d'état in some of the smaller unstable republics." In addition, climate change has become a political issue, which can color how new observations are reported.

If I do a good job of choosing and explaining the quantities that are needed to define and understand climate change, it may help you as you try to assess the continuing course of Earth's climate, the forces that drive climate change, and the impacts of climate change. Quantities selected must be not only central to the "physics" of the problem but also measurable with a meaningfully small uncertainty.

In summary, the two key quantities for the carbon cycle are the annual carbon dioxide emissions from fossil fuel burning and the annual increase of atmospheric carbon dioxide. Regarding the first quantity, note in figure 16 that the Kyoto Protocol, which was adopted in 1997 and went into force in 2005 with "legally binding commitments" to reduce greenhouse gas emissions, did not lead to a decrease in global emissions—indeed, emissions continued to increase. The second quantity, annual change of carbon dioxide in the air (figure 17), so far is closely following business-as-usual scenarios, which are based on the assumption that all fossil fuels will be burned. Finally, the ratio of these two key quantities, the dashed curve in figure 16, provides an additional important conclusion: The net "sink" for human-made carbon dioxide is not decreasing; rather, the sink is increasing, as it continues to average about 44 percent of emissions, which are increasing.

It would be difficult to exaggerate the importance of tracking these quantities. Indeed, continuation of life on this planet requires a rapid change of the trajectory of these quantities.

NOW BACK TO the Keeling talk and its repercussions. There was no press release or press conference about the talk, but the American Geophysical Union meeting attracts a substantial number of reporters. BBC radio did an impromptu interview with me as I left the speaker's platform. Bill Blakemore used a quote from my talk in an ABC News story the next day. The *New York Times* and the *Washington Post*, in articles about international climate negotiations, made note of my comment that 2005 was likely to be at least as warm as 1998, the previous warmest year in the period of instrumental data. The *International Herald Tribune* extracted several paragraphs from my talk, verbatim, making a short article under my byline.

Unbeknownst to me, this modest level of publicity was causing growing concern in the Office of Public Affairs at NASA headquarters. And the next week, on December 15, this festering consternation of NASA officials exploded into what the agency's public affairs employees described as a "shitstorm." The immediate cause of the explosion was the statement on ABC's *Good Morning America* program that "NASA is announcing that this year, 2005, is tied for the hottest year ever." ABC did not mention my name, but indeed I had provided our analysis of global temperature for the meteorological year (December through November) to Bill Blakemore the previous day.

The story of this storm reported in the press more than a month later, most prominently in a front-page article by Andrew Revkin in the January 29 *New York Times*, focused on a twenty-four-year-old political appointee in the Office of Public Affairs. The impression left with the public was that this young man in the NASA press office had taken it upon himself to censor what I was allowed to say in public. Congress, despite available information to the contrary, decided to publicly accept this low-level maverick story, which NASA management foisted on it. As a result, fundamental problems for the functioning of our democracy were left unexamined.

In reality, the highest levels at NASA headquarters were involved in the censorship, and their clumsy attempts at silencing me were instigated in response to calls from the White House. The facts became clear only with the investigative reporting of Mark Bowen for his book *Censoring Science*, through a NASA Office of Inspector General investigation and report published several months after Bowen's book, and by a penetrating ten-page summary, including new information, posted November 18, 2008, by Bowen on his blog Tipping Points. But I'd like to describe my view of the storm, including some information I learned through Bowen's reporting, because there are broader implications that go far beyond the little tempest in NASA.

Of course by 2005 I was well aware that the NASA Office of Public Affairs had become an office of propaganda. In 2004 I learned that NASA press releases related to global warming were sent to the White House, where they were edited to appear less serious or discarded entirely. In connection with my University of Iowa talk in 2004, I had informed the *New York Times* and National Public Radio about this practice, as well as the fact that NASA's head of public affairs, Glenn Mahone, had driven to the Goddard Space Flight Center in Greenbelt, Maryland, to verbally chew out the young contractors who had informed me about the role of the White House. I attributed the fact that both the *Times* and NPR did not report this story to their concern about the possibility of a lawsuit. Although there were witnesses to the "chewing out" administered by Mahone, there was no paper trail confirming that draft press releases were being passed to the White House.

In any case, on the morning of December 15, I was working hard in my apartment against a self-imposed deadline for my paper on dangerous anthropogenic climate change, which I wanted to submit for publication by the end of 2005.

So when Larry Travis, my deputy, and Leslie McCarthy, the NASA public affairs officer in New York, called me that morning to tell me that a "shitstorm" was under way at NASA headquarters, that I should expect an admonishing phone call from the associate administrator for science, and that strict procedures were to be implemented to prevent me from speaking with the media, my first reaction was to laugh and say, "What else is new?" then get back to work on my paper. But I soon realized that the situation was different this time. No longer would the Office of Public Affairs simply be failing to help publicize research results; now it would be clamping down on communication with an iron grip.

The first clampdown occurred that very afternoon. NASA headquarters ordered the removal of our global temperature analysis from the Goddard Institute for Space Studies Web site. Although I am the NASA official in charge of the GISS Web site, the order did not come through me. Instead, the NASA Public Affairs Office directly instructed the GISS webmaster, who is a contract employee, to remove the analysis. The clout of public affairs employees was further demonstrated that day as they enlisted both the director of the Goddard Space Flight Center, Ed Weiler, and the NASA associate administrator for science, Mary Cleave, to personally review the page that I had written for the Web site to accompany our global temperature analysis. Both officials agreed that the write-up was good science and that our temperature analysis and its description should be put back on the GISS Web page. However, Public Affairs overruled the NASA associate administrator and kept our data analysis off the Web site until the next day.

On December 16, I was informed of the rules that I must henceforth live under. These rules had been laid out in a late-afternoon telephone call the previous day from NASA headquarters Public Affairs in Washington to Leslie McCarthy in New York.

As reported in Bowen's book, *Censoring Science*, Leslie McCarthy was the only person on the New York end of the telephone line. Leslie does not work for NASA headquarters. She is an employee of the Goddard Space Flight Center's Public Affairs Office in Greenbelt, Maryland, assigned to the Goddard Institute for Space Studies, the New York division of Goddard. She reported to Mark Hess, the head of Goddard Public Affairs. The call was scheduled to take place after normal work hours, and Leslie anticipated that it would be difficult. She requested that a representative of the Goddard Public Affairs Office be on the line during the conversation, but headquarters would not allow her to bring anyone from Greenbelt into the conversation.

Four people were on the Washington end of the line: David Mould, Dean Acosta, Jason Sharp, and George Deutsch—all political appointees of the Bush administration. They were in Dean Acosta's office using his speakerphone. McCarthy was never informed of the presence of Sharp and Deutsch, who did not speak during the call.

Partway through the conversation, the four were joined by Dwayne Brown, a NASA career public affairs officer.

David Mould, who had succeeded Glenn Mahone as head of public affairs for NASA, had been at the agency only six months; his title was assistant administrator for public affairs. As Bowen reports in *Censoring Science*, during George W. Bush's 2000 presidential campaign, Mould held senior positions in public and media relations at the Southern Company of Atlanta, the second-largest holding company of coal-burning utilities in the United States and thus the second greatest emitter of carbon dioxide. Southern's contributions to the Republican Party were exceeded that year only by Enron's.

Dean Acosta was second in command to Mould and also the press secretary for the NASA administrator. Acosta had been at NASA since 2003, thus providing Public Affairs with an institutional familiarity and memory. Bowen suggests that this is the reason why Acosta received one of the phone calls from the White House on December 15 regarding the ABC report.

Jason Sharp was Mould's assistant. George Deutsch was a twenty-four-year-old presidential appointee, who, according to his résumé, had a bachelor's degree in journalism from Texas A&M University. He had worked as an intern in the "war room" of the Bush-Cheney reelection campaign and had been hired by NASA on the recommendation of Acosta. Deutsch's first supervisor, Dolores Beasley, told Mark Bowen that she had to correct Deutsch more than once for saying that his job at NASA was "to make the president look good."

The phone conversation began with Dean Acosta expressing their great consternation about having been "blindsided" by the ABC report. But the primary purpose of the call was to describe new "rules of engagement." These rules, they said, would apply to anyone with a NASA badge or funded by NASA, but they made clear that I was their primary concern. David Mould declared that under the new policy from NASA administrator Michael Griffin (Sean O'Keefe's successor), no one was to take a direct call from the media without notifying the Public Affairs Office. Mould said that this rule was being put in place directly by the administrator, and the specific procedures would be spelled out by Mary Cleave. Mould said that they were "tired of Jim Hansen trying to run an independent press operation." From then on, they wanted to know everything I was doing or planning to do, and said that Leslie must keep them informed, well ahead of time, of every item on my calendar that might attract media attention.

During the conversation Dwayne Brown entered the room and
described two of the new rules of engagement, based on instructions he had just received from Mary Cleave. First, all content
posted on our Web page would require prior approval. Even papers
that had been accepted for publication in scientific journals could
not be posted until they were explicitly approved by NASA headquarters. Second, all requests for interviews must be forwarded to
headquarters, where Cleave and her deputy, Colleen Hartman,
would have the "right of first refusal" on all interview requests.
That is, they would be interviewed themselves, unless they deferred, in which case they would suggest the most appropriate person to be interviewed.

The following day Leslie McCarthy's boss, Mark Hess, who was
on travel at another Goddard division on Wallops Island, Virginia,
received a request to call Dean Acosta. David Mould joined in the
resulting conversation, in which they repeated to Hess the same instructions they had given Leslie McCarthy.

Hess and McCarthy had misgivings about the propriety of these
new rules. They decided to delineate the rules on paper and ask
Mould and Acosta to verify their accuracy. According to these rules,
reporters would be allowed access to NASA scientists only through
Public Affairs. All material on NASA Web sites required prior notification and approval of headquarters, including Public Affairs. Any
activity, speech, or data release that might generate media attention
must be reported to Public Affairs well ahead of time.

The teeth in the new rules were tested within days. National
Public Radio requested an interview with me for its *On Point* program. Deutsch, as was his wont, scampered straight to the ninth
floor, where the offices of the administrator and his senior managers, including Mould and Acosta, are located. Deutsch then informed McCarthy that "the ninth floor" did not want me to appear
on NPR because it was "the most liberal news outlet in the country." To guard against the possibility that I might agree to the interview, Dwayne Brown called McCarthy to say that there would be
"dire consequences" if I appeared on the program. The request was
diverted to Cleave and Hartman, but NPR staff decided not to go

through with the interview, because they wanted to speak specifically with me.

I learned that another interview request, from the *Los Angeles Times* for information on our 2005 temperature analysis, had been diverted by Deutsch to a Goddard scientist who had no knowledge of global temperature data. So it was clear I could not just ignore what the Public Affairs Office was doing. I needed to fight back if I wanted to retain an ability to communicate with the outside world.

But, for the time being, I had to finish the "Dangerous" paper, which was finally submitted to the *Journal of Geophysical Research* on December 30. I included a letter to the editor apologizing for the paper's great length. I noted that it could be broken into two or three papers, but that would only increase the total length. I requested that he consider publishing it as a single long paper, so as to retain its coherence.

I then grumpily began to complete mandatory annual government "training" exercises, which are meant as a reminder of regulations on NASA standards of conduct, ethics, equal opportunity, and so on. It was a waste of time, I thought, to keep repeating this every year. Then it dawned on me that words I was reading provided arguments that I could use for fighting the censors. The first line of NASA's mission statement: "to understand and protect our home planet." The last word in NASA's core values: "integrity"—defined as honesty, ethical behavior, respect, candor. The first principles of government ethics: "Public service is a public trust."

Stitching these principles together with the evidence for dangerous human-made climate change, I wrote a memo to higher levels of Goddard management arguing that we had to fight the Public Affairs restrictions. I concluded the memo by saying, "If NASA is to fulfill its mission of providing information that helps the public and policymakers understand and protect our home planet, if it is to uphold its public trust with integrity, it cannot knuckle under to political pressures."

Of course Goddard management was not the problem. The problem was at NASA headquarters. The censors were at the right hand of the administrator, operating with his approval. Goddard could not do much to affect these higher levels. But the memo, and its attachments explaining the climate threat, would be useful for informing others about the situation.

On the morning I finished and e-mailed my memo to Goddard management, I also was preparing for an afternoon interview with Scott Pelley for the CBS program *60 Minutes*. The interview, expected to focus on Arctic climate change, had been scheduled several months earlier; headquarters had been informed and had no basis to prohibit it. I sent the producer, Catherine Herrick, information on the recent censorship by NASA headquarters and suggested that if NASA insisted on sending a public affairs officer to be with me during the interview, that the network pan the camera to show the NASA "minder."

I felt awkward during the interview, as usual, and could not remember the exact quote that I wanted to use: "In my more than three decades in the government, I have never seen anything approaching the degree to which information flow from scientists to the public has been screened and controlled as it is now"—a statement I had made to the *New York Times* that was reprinted on the Freedom Forum calendar. The producer did a good job, though. The program demonstrated that the administration was editing climate information, downgrading the implications, and showed specific edits that had been made to climate reports. It closed with the comment that the "editor" had just left the White House for a job with ExxonMobil.

However, such an exposé, no matter how compelling, does little to cure the cancer that afflicts communication of scientific information to the public. Neither political party, I will argue, has been willing to fix this problem, which is a threat to democracy and humanity.

I had another opportunity to communicate the censorship matter the next day, when I had lunch with Al Gore. It was my first meeting with Gore in more than a decade. I had fallen out of favor with the Clinton-Gore White House after declining to write a rebuttal to an op-ed in the *New York Times* that criticized Gore's views on climate change. At the time, that was fine with me; I preferred to be left to do research.

When I met Gore, he feigned being miffed that I did not invite him to visit the GISS office. "What's the matter, am I radioactive?"

"Well, yeah, you probably are," I said, laughing. A visit by Al Gore to our laboratory might have sent my superiors into a tizzy, especially since I had not informed anyone about the meeting. And, despite ongoing difficulties with the Bush-Cheney administration, I did not want to be identified with either political party.

Lunch was at the Regency Hotel on Park Avenue. I agreed to one of his requests, to review critically the science in his slide show on global warming, and demurred on another, to be on a board overseeing a media campaign to inform the public about global warming. As we got up to leave, he introduced me to two people at a nearby table: Larry King and Norman Pearlstine, mentioning to them my current travails with Public Affairs censorship.

Larry King could have provided an opportunity to inform the public about climate change and the censorship issue. I told him that censorship was at least as bad at EPA and NOAA as at NASA, and that the practice would surely have a negative effect on national decision making. King was sympathetic, but upon hearing that the major impacts of current bad policy would occur several decades in the future, he declared, "Nobody cares about fifty years from now."

Norman Pearlstine, who had just stepped down as editor in chief at Time Inc., and maintained close connections there, was more interested. He asked if I would be willing to go public, to describe what was happening on the record. I agreed, and said that I would provide a detailed description of the situation.

I spent the weekend writing a comprehensive discussion of the events of the past two years, the science, and the implications. I emphasized the danger of passing climate tipping points and the urgency of policy actions. I suggested that a reason for censoring me was that I had begun connecting the dots all the way from the science to needed policy actions, including a discussion of the reasons that these actions were not being taken.

Anniek hand-delivered the package—including copies of the "Iowa" and "Keeling" talks and the "Dangerous" paper—to Pearlstine. A week or so later, after hearing nothing from *Time*, I sent essentially the same material to Andrew Revkin of the *New York Times*.

Pearlstine passed the material down the line at *Time*. Eventually I received a call from one of its writers, suggesting an interview sometime in the future for a special issue on global warming.

Fortunately, Andy Revkin was both interested and a sharp investigator. It is not easy to get the approval of editors for an "accusatory" article. According to Revkin, the key factor was the willingness of career civil servants Leslie McCarthy and Larry Travis to go on the record with concrete information. McCarthy had detailed notes of correspondence with the NASA headquarters Public Affairs Office and provided specific quotes to Revkin.

The article, "Climate Expert Says NASA Tried to Silence Him," which appeared in the top right-hand column on the front page of the Sunday, January 29, 2006, *New York Times*, got immediate attention and seemed like it could lead to some real good. Sherwood Boehlert, the Republican chairman of the House Committee on Science and Technology, told his chief of staff, David Goldston, that he wanted "to do something on this right away."

Boehlert was an outstanding representative and one of the best friends that science has ever had in Congress. Thus on October 11, 2006, at a meeting of the League of Conservation Voters, where Boehlert and I were both speakers, my jaw surely dropped when I heard him declare that the affair at NASA had been entirely the work of a renegade twenty-four-year-old and that Administrator Griffin had fixed the problem comprehensively.

After Boehlert stepped down from the speaker's platform, and as people were milling around, I approached him and said, "What you said isn't right. The problem went all the way to the top, and it hasn't been fixed."

Boehlert put his hand on my shoulder and said, "I know, I know." But he had just said the opposite to a room of several hundred people. Perhaps that's the way it works in Washington. People learn to lie with a straight face. Even the good guys.

That was my immediate reaction, but such an interpretation is inconsistent with the fact that Boehlert, now retired, was and is universally respected, even loved, by colleagues and constituents. His service to the country and his integrity are above reproach. Boehlert was properly offended by the evidence of censorship of science, and he made clear to Administrator Griffin that the members of the House Committee on Science and Technology would not tolerate scientific muzzling or censorship and that they would be watching to make sure that NASA corrected any problems that existed.

Mark Bowen writes in *Censoring Science* that David Goldston told him that Boehlert "was relieved that there was such a tidy story . . . There was this rogue guy [Deutsch] . . . and they got rid of him." Bowen summarized the matter thusly: "Goldston thinks Boehlert truly believed the tidy story, even though he was told 'millions of times' that there was more to it than that. And the chairman's public remarks from that time forward would reflect his tidy belief."

As for me, I interpret Boehlert's "I know, I know" and pat on the shoulder as being a general expression of support or sympathy. He probably did not hear exactly what I said.

Fortunately, at about the same time that Boehlert was expressing his opinion about the narrow responsibility for censorship, fourteen U.S. senators cosigned a letter to the NASA inspector general asking him to conduct an investigation into the allegations of "political interference" with scientists at NASA. Unfortunately, Inspector General Robert Cobb hardly seemed the ideal person for such an investigation.

The Associated Press reported in April 2007 that e-mails revealed that Cobb had met regularly with NASA administrator Sean O'Keefe, played golf with him, and tipped him off about impending audits. The President's Council on Integrity and Efficiency revealed that Cobb had quashed a report on the *Columbia* shuttle disaster that would have embarrassed NASA. Bowen discusses these matters and the fact that many NASA employees were afraid to communicate with the inspector general because of his perceived cozy relationship with managers from whom he was supposed to be independent.

Nevertheless, I was convinced that the members of the Inspector General's Office who contacted me were genuinely interested in getting at the truth, and I believe that they did a good job of collecting information. They suggested that the investigation would probably take a few months. For some reason, it took almost two years. The inspector general was made aware of Bowen's investigations and book in preparation. It did not surprise me that the inspector general's report did not come out until June 2008, several months after Bowen's book.

The report, titled "Regarding Allegations that NASA Suppressed Climate Change Science and Denied Media Access to Dr. James E. Hansen, a NASA Scientist," confirmed the allegations and placed the blame several layers higher than suggested by the *New York Times* article and Congressman Boehlert: on the leadership of the NASA Public Affairs Office, Mould and Acosta. The report concluded that the Public Affairs Office's actions in editing and downgrading press releases and denying media access were inconsistent with NASA's obligation under the National Aeronautics and Space Act to achieve the widest practical dissemination of information concerning its activities and results.

The report places blame squarely on public affairs employees, absolving the NASA administrator of responsibility for their *unilateral* (the inspector general's emphasis) actions. The investigators also were unable to confirm that administration officials outside NASA approved, disapproved, or edited news releases. However, they concluded that "the preponderance of evidence supported claims of inappropriate political interference in dissemination of NASA scientific results."

Bowen does not agree with the absolution of NASA administrator Michael Griffin or the absence of White House involvement, as he explained in a November 18, 2008, post on his blog, cross-posted on Daily Kos. Bowen points out that the instructions to Leslie McCarthy on December 15, 2005, were described as emanating from Griffin. Mary Cleave on that day told Dwayne Brown that Griffin had received one of the calls from the White House. Also, J. T. Jezierski, Griffin's deputy chief of staff and White House liaison, told Bowen that on December 15 he had received an angry call from the White House and added that "the 'sustained media presence . . . of Dr. Hansen' was the dominant issue all that day and the next for every top official in public affairs and communications at the agency—himself, chief of staff Paul Morrell, strategic communications director Joe Davis, and David Mould—and that these officials also held discussions with Michael Griffin during those two days."

Bowen suggested holding an inquiry that calls the senior players, including Griffin, to testify under oath and the threat of perjury but without fear of retribution. Perhaps there is merit in that. But I believe that the fundamental source of the problems is clear, and it could be readily fixed. Unless it is resolved, the problems will surely recur in the future, even if carefully camouflaged.

The key matter is hinted at in the inspector general's report, which notes that the political appointees running the Office of

Public Affairs had the "seemingly contradictory position" of being expected to ensure the "widest practicable dissemination" of research results that, in this case, were inconsistent with the administration's policies. Most political appointees are smart enough not to say "my job is to make the president look good," but they know what their job is. The report suggests no remedy for their contradictory position. Its main effect is to remind the appointees to avoid slipping up, by leaving paper trails, for example.

In my talks I began to emphasize the first line of the NASA mission statement: "to understand and protect our home planet." This perspective had removed any doubt in my mind about whether it was appropriate to connect all the dots. Andy Revkin reported, in his front-page *Times* article, that it was because of this mission objective that I had decided it "would be irresponsible not to speak out."

The NASA mission statement had been arrived at years earlier, under administrator Dan Goldin, with great fanfare. A committee was set up at headquarters, suggestions were sought, and there were iterations with all NASA employees. The resulting mission statement, almost everyone agreed, was inspiring.

But in the spring of 2006, a NASA colleague sent me an e-mail warning me that I had better stop using that statement as a rationalization for my actions because it no longer existed. Sure enough, when I checked the mission statement on the NASA Web site, the phrase "to understand and protect our home planet" was gone. I checked with more than a score of people at headquarters, Goddard management, scientists, and people at other NASA centers, and nobody knew what had happened. It had just disappeared.

Another disappearance occurred simultaneously, almost as if two mirror particles, matter and antimatter, had collided, and *poof*, both were gone. The second thing to disappear was 20 percent of the NASA earth science research and analysis budget. An earth science manager called me to tell me of the cut and the hard times ahead. Because most of the budget goes toward fixed items, such as rent and civil service salaries, a 20 percent cut is monstrous, a signal almost of going out of business.

The Constitution grants Congress the power of the purse strings. Although the executive branch has found ways to gradually assume increasing authority, Congress never would have acceded, knowingly, to the decimation of earth science research and analysis. But the budget cut was inserted via a clever stealth maneuver. When the proposed budgets for the upcoming year were sent to Congress and reported in the media, NASA earth science had a change of only a percent or two, typical of other programs and nothing that would raise any eyebrows.

Here was the trick. In a little-noticed "operating plan" for the current year, which the administration submitted to Congress just prior to the budget for the upcoming year, there was a cut of about 20 percent in planned expenditures for earth science research and analysis retroactive to the beginning of the current year. More than a third of the way into the fiscal year, NASA earth science managers were told to rebalance their books with a draconian cut.

It turns out that the change to NASA's mission statement was slipped into this same operating plan. A change in the budget inserted in the operating plan is something that would have been worked out between the White House Office of Management and Budget and the NASA Administrator's Office. As for the removal of the mission statement, Mark Bowen reported in his book: "A high insider at headquarters told me that Michael Griffin rewrote the mission statement and the agency's strategic plan basically on his own."

For the purpose of drawing attention to these maneuvers, I wrote an article, "Swift Boating, Stealth Budgeting, and Unitary Executives," that was published in *World Watch* magazine. My hope was that, by exposing these unsavory deeds, I might stir someone in Congress to take offense at the grasping by the executive branch, perhaps even to exercise his or her constitutional prerogative. So I ended my article on a "hopeful" note:

> But may it be that this is all a bad dream? I will stand accused of being as wistful as the boy who cried out, "Joe, say it ain't so!" to the fallen Shoeless Joe Jackson of the 1919 Chicago Black Sox, yet I maintain the hope that NASA's dismissal of "home planet" is not a case of either shooting the messenger or a too-small growth of the total NASA budget, but simply an error of transcription. Those who have labored in the humid, murky environs of Washington are aware of the unappetizing forms of life that abound there. Perhaps the NASA playbook was left open late one day, and by chance the line "to understand and protect our

home planet" was erased by the slimy belly of a slug crawling in the night. For the sake of our children and grandchildren, let us pray that this is the true explanation for
the devious loss, and that our home planet's rightful place
in NASA's mission will be restored.

Protection of our home planet, I suggest, is intimately related to
protection of our democracy. The American Revolution launched
the radical proposition that the commonest of men should have a
vote equal to that of the richest, most powerful citizen. Our forefathers devised a remarkable Constitution, with checks and balances,
to guard against the return of despotic governance and subversion of
the democratic principle for the sake of the powerful few with special interests. They were well aware of the difficulties that would be
faced, however, placing their hopes in the presumption of an educated and honestly informed citizenry.

I have sometimes wondered how our forefathers would view our
situation today. On the positive side, as a scientist, I like to imagine how Benjamin Franklin would view the capabilities we have
built for scientific investigation. Franklin speculated that an atmospheric "dry fog" produced by a large volcano had reduced the
sun's heating of Earth, causing unusually cold weather in the early
1780s; he noted that the enfeebled solar rays, when collected in the
focus of a "burning glass," could "scarce kindle brown paper." As
brilliant as Franklin's insights may have been, they were only speculation, as he lacked the tools for quantitative investigation. No
doubt Franklin would marvel at the capabilities provided by Earth-
encircling satellites and supercomputers that he could scarcely
have imagined.

Yet Franklin, Jefferson, and the other revolutionaries would surely
be distraught by recent tendencies in America, specifically the increasing power of special interests in our government, concerted
efforts to deceive the public, and arbitrary actions of government
executives that arise from increasing concentration of authority in
a unitary executive, in defiance of the aims of our Constitution's
framers.

I believe there is a straightforward way to address these issues. I
made some suggestions before a congressional committee with a
name that sounded promising: the House Committee on Oversight

and Government Reform. I thought my suggestions were substantive. I was disappointed that the committee members basically ignored them. Their interest seemed to be partisan posturing, not reform or solutions. The public needs to be aware of these matters and put pressure on government to fix the problems. Here are my suggestions.

First, abolish the practice of placing political appointees in Public Affairs Offices of the science agencies. Science and its reporting to the public should not be political. Public Affairs Offices should be staffed by career professionals protected from political pressures by civil service regulations. What's so hard about that? If political appointees are placed in Public Affairs it is prima facie evidence that the administration wants an office of propaganda.

President Barack Obama has drawn attention to the censoring of science that occurred under the Bush administration and has promised to correct the problem. However, if political appointees are still placed in the Public Affairs Offices of the science agencies, any paper changes in the rules or safeguards are practically meaningless. In my experience, there is no qualitative difference between Democratic and Republican administrations. Public Affairs does fine on most scientific results, because most do not have political overtones. But on a sensitive topic such as global warming, there is interference. The most political interference that I had on a press release was late in the Clinton-Gore administration, with our alternative scenario paper, which tried to draw attention to the importance of climate forcings other than carbon dioxide.

The Public Affairs problem could be fixed by a law that prohibited political appointees in those offices. Or the president could fix it immediately, by inserting no political appointees into Public Affairs. If a later president reinserted political appointees, the public could be promptly informed that these offices were again functioning as propaganda offices.

Second, abolish the requirement for government scientists to have their testimony to Congress reviewed and edited by the White House Office of Management and Budget. Government scientists do not work for the president; they work for the taxpayer. Congress and the public have the right to hear unfiltered testimony. This censorship has no basis in the Constitution. It is just one of the ways that the executive branch has arbitrarily increased its power.

If the president continues this practice, Congress should vociferously object, taking the president to court, if necessary.

However, the biggest obstacle to solving global warming is much more difficult than the two specific matters above. The problem concerns the role of money in politics, the undue sway of special interests. But before we discuss this crucial issue, I need to give you some bad news. The dangerous threshold of greenhouse gases is actually lower than what we told you a few years ago. Sorry about that mistake. It does not always work that way. Sometimes our estimates are off in the other direction, and the problem is not as bad as we thought. Not this time. The bad news emerged clearly only in the past three years. That is the story in the next chapter.

CHAPTER 8

Target Carbon Dioxide: Where Should Humanity Aim?

In 2007, THE ENVIRONMENTALIST AND writer Bill McKibben began bugging me, very politely, to either confirm 450 parts per million as the appropriate target level of carbon dioxide in the atmosphere or else to define a more appropriate one. He was developing a Web site to draw attention to this target limit and was thinking of calling it 450.org. I kept putting him off, though. I wanted a number that would remain valid for policy purposes for the foreseeable future. And I wanted to have a good science rationale—otherwise the number would have little meaning.

The issue of what is the "dangerous" level of greenhouse gases has been around a long time. Back in 1981 my coauthors and I concluded (in our paper "Climate Impact of Increasing Atmospheric Carbon Dioxide," published in *Science*) that serious effects of climate change might make it necessary to leave a large part of the coal in the ground. And in 1992 most countries of the world, including the United States, signed the United Nations Framework Convention on Climate Change, with the objective of stabilizing atmospheric greenhouse gases at a level that would avoid "dangerous" climate change.

By the late 1990s I had begun to work explicitly on the question of what the dangerous level would be. I also wanted to better understand the degree to which there could be a trade-off ("offset") between carbon dioxide and other human-made climate forcings.

The result of this research was our "alternative scenario" paper, published in 2000. We concluded that carbon dioxide had better be kept to no more than about 450 ppm. And it should be allowed to go that high only if some other gases, notably methane and tropospheric ozone, were reduced below current values, a task that we argued was feasible but not easy. A target of 450 ppm would mean 1 degree Celsius additional warming.

Where did this limit of 450 ppm come from? Not from climate models, although it is easy to understand why the public would believe that—they hear it again and again, from people with a vested interest in covering up the reality and urgency of the climate threat. They know, just as Nazi propaganda chief Joseph Goebbels did, that if you repeat something often enough, many people will believe it. And they know that it is easy to find fault with climate models, which are still very imperfect representations of the real world. So they set up a strawman—pretending that the 450 ppm limit came from climate models. Actually the 450 ppm limit came from looking at Earth's history—remarkably detailed data showing how Earth responded in the past to changes of climate forcings, including changes of atmospheric composition.

Civilization developed in and is adapted to the climate of the Holocene, the stable, relatively warm period that has existed for about 11,000 years. There have been prior interglacial periods—warm intervals between the ice ages—during the past several hundred thousand years that were warmer than the Holocene; as much as a few degrees Celsius warmer at the poles, but only about 1 degree warmer on global average.

My thesis was that Earth during those interglacial periods was reasonably similar to Earth today. On the other hand, if we go back to the last time that Earth was 2 or 3 degrees warmer than today, which means the Middle Pliocene, about three million years ago, it was a rather different planet. Sea level was about 25 meters (80 feet) higher than today. Florida was under water. About a billion people now live at elevations less than 25 meters. It may take a long time for such large a sea level rise to be completed—but if we are foolish enough to start the planet down that road, ice sheet disintegration likely will continue out of our control.

The target limit of 1 degree warming that we arrived at is relative to the temperature in 2000. Earth warmed about 0.7 degree Celsius between the 1800s and 2000, which made the global temperature in 2000 approximately match the highest level in the Holocene. So the limit on warming that we suggested was 1.7 degrees Celsius relative to the late Holocene, the preindustrial climate, just prior to the warming that began in the late 1800s. Our proposed limit on global warming was only a few tenths of a degree stricter than the 2-degree limit the European Union has been advocating for the past few years and continues to advocate today.

Unfortunately, what has since become clear is that a 2-degree Celsius global warming, or even a 1.7-degree warming, is a disaster scenario. In order to clarify why I can say that with confidence, I need to continue the story of what has transpired in the past several years.

Even when I first suggested a limit of 1 degree Celsius additional warming, and a carbon dioxide limit of about 450 ppm, I was aware of one niggling detail—but I swept it under the rug. That niggling detail was the evidence that sea level during the prior interglacial period, about 125,000 years ago, had probably reached a level about 4 to 6 meters higher than today. A sea level rise of 5 meters (about 17 feet) would submerge most of Florida, Bangladesh, the European lowlands, and an almost uncountable number of coastal cities around the world.

My rationale for sweeping my concerns about sea level under the rug was the belief that ice sheets, and thus sea level, can change only very slowly. If that were true, then humanity might have a thousand years to figure out how to get atmospheric composition back to a safe level or adapt to changing sea level. That tidy rationalization seemed to be supported by ice sheet models and by most paleoclimate records of sea level change.

Unfortunately, over the past several years, support has crumbled for the tidy belief that ice sheets require millennia to disintegrate.

First, as I argued in my 2005 "Slippery Slope" paper and discussed in chapter 5, it became clear that the ice sheet models fail to incorporate physics components that are critical during ice sheet collapse. This deficiency has been confirmed by the models' inability to simulate the rapid changes observed on Greenland and Antarctica during the past few years.

Second, the belief that ice sheets are inherently lethargic is based mainly on the average rate at which they grew and decayed during Earth's history. The overall size of ice sheets grew and decayed over tens of thousands of years. But the ice sheets responded so slowly because that was the time scale for changes of Earth's orbit—the time scale for the forcings that caused ice sheets to grow or melt. Those slow orbital changes imply nothing about how fast the ice sheets would respond to a rapid forcing. On the contrary, as I and five coauthors showed in a paper published in the *Philosophical Transactions of the Royal Society* in 2007, during the last deglaciation there was no discernible lag between the time of maximum solar forcing of the ice sheet and the maximum rate of melt (maximum rate of sea level rise). In other words, paleoclimate data indicate that ice sheets are able to respond rapidly, with large changes within a century. Sea level 13,000 to 14,000 years ago rose at a rate of 3 to 5 meters (10 to 17 feet) per century for several centuries.

Third, evidence has mounted during the past several years that it is not unusual for sea level to fluctuate by several meters within an interglacial period. The most comprehensive study for the immediately prior interglacial period was published by geologist Paul Hearty and several colleagues in 2007. They showed, from sedimentary and fossil evidence on the shorelines of Australia, Bermuda, the Bahamas, and other locations, that sea level was about 2 meters above the present level for most of that interglacial period, but near the end of it, about 120,000 years ago, sea level increased to a maximum between 6 and 9 meters higher than today. The additional water must have come from Antarctica, Greenland, or some combination of the two.

The Hearty study and others show that the sea level stability of the late Holocene cannot be taken for granted. The Holocene's stable sea level, so far, may be related to the fact that temperature peaked early in the Holocene and at a level slightly cooler than in most interglacial periods—and this peak warming was followed by a slight cooling trend.

Whatever the reason for sea level stability, it helped spur the development of civilization, as mentioned earlier. The stable sea level not only provided early humans with a high-protein marine food supply, but it also made possible grain production in estuary and floodplain ecosystems. With these conditions, food for the human population could be produced by a fraction of the people, thus allowing a transition from the Neolithic way of life to urban social life and the development of complex state-governed societies.

The period of stable sea level is almost surely over. But whether human-caused sea level rise will be a slow bump-up reaching a maximum only on the order of a meter or so, or whether it will be an eventual increase of tens of meters, with disintegrating ice sheets, continual havoc for coastal cities, and a redrawing of global coastlines, depends on policies adopted in the near term. I believe it is possible to keep sea level rise at a small bump-up, but that will require the amount of atmospheric carbon dioxide to peak soon and then begin at least a moderate decline.

Sea level rise is one of the two climate impacts that I believe should be at the top of the list that defines what is "dangerous," because the effects are so large and because it would be irreversible on any time scale that humanity can imagine. Ice sheets take thousands of years to build up from snowfall. Reasonable "adaptation" to a large sea level rise is nearly impossible, because once ice sheets begin to rapidly disintegrate, sea level would be continually changing for centuries. Coastal cities would become impractical to maintain.

The other climate change impact at the top of my "dangerous" list is extermination of species. Human activities already have increased the rate of species extinctions far above the natural level. Extinctions are occurring as humans take over more and more of the habitat of animal and plant species. We deforest large regions, replace biologically diverse grasslands and forests with monoculture crops, and introduce foreign, invasive animal and plant species that sometimes wipe out the native ones.

Now human-made climate change, with an unnaturally rapid shifting of climatic zones, threatens to add a new overwhelming stress that could drive a large fraction of the species on the planet to extinction. Our understanding of this threat, as in the case of ice sheets and sea level, depends especially on information that we extract from Earth's history and observations of what is happening today. There is another analogy between sea level change and species extermination: Survival of ice sheets and species both present "nonlinear" problems—there is a danger that a tipping point can be passed, after which the dynamics of the system take over, with rapid changes that are out of humanity's control.

In 2006, after a television appearance in which I criticized the Bush administration's censorship of science, I received the following e-mail from a man in northeast Arkansas: "I enjoyed your report on *60 Minutes* and commend your strength. I would like to tell you of an observation I have made. It is the armadillo. I had not seen one of those animals my entire life, until the last ten years. I drive the same 40-mile trip on the same road and have slowly watched these critters advance further north every year and they are not stopping. Every year they move several miles."

Indeed, animals are on the run. Plants are migrating too. Earth's creatures, save for one species, do not have thermostats in their living rooms that they can adjust for an optimum environment. Animals and plants are adapted to specific climate zones, and they can survive only when they are within those zones. In fact, scientists often define climate zones by the vegetation and animal life that they support. Gardeners and bird-watchers are well aware of this, and their handbooks contain maps of the zones in which a tree or flower can survive and the range of each bird species.

Those maps are already being redrawn. Most people, mainly aware of much larger day-to-day fluctuations in the weather, hardly notice that climate, the average weather, is changing. In the 1980s I used a colored six-sided die that I hoped would help people understand global warming at an early stage. Only two sides of the die were red, or hot, representing the probability of having an unusually warm season during the years between 1951 and 1980. By the first decade of the twenty-first century, four sides were red; one side was white, for average 1951–1980 conditions; and one side was blue, for unusually cold. The actual change in the frequencies have almost matched that expectation—the number of unusually warm seasons now averages about 60 percent, while it needs to reach 67 percent to yield four red sides.

As climate change continues, species must migrate to survive. I would not worry too much about the armadillos. Although their ingenuity may be taxed a bit, as they seek ways to ford rivers and multilane highways, they are tough, mobile critters that have survived climate changes for more than 50 million years. Compare that with the puny 200,000 years that *Homo sapiens* have existed, or even the 2 million years of our predecessor, *Homo erectus*.

Problems are greater for other species. Ecosystems are based on interdependencies—between, for example, flower and pollinator, hunter and hunted, grazer and plant life—so the less mobile species have an impact on the survival of others. Of course, species adapted and flourished during past climate fluctuations. But now the rate of climate change driven by human activity is reaching a level that dwarfs natural rates of change. And barriers created by human beings, such as urban sprawl and homogeneous agricultural fields, block many migration routes. If climate change is too great, natural barriers, such as coastlines, will spell doom for some species.

Studies of more than one thousand species of plants, animals, and insects (including butterfly ranges charted by members of the public) found an average migration rate toward the north and south poles of about four miles per decade in the second half of the twentieth century. That is not fast enough. During the past thirty years the lines marking the regions in which a given average temperature prevails ("isotherms") have been moving poleward at a rate of about thirty-five miles per decade. That is about the size of a county in Iowa. Each decade the range of a given species is moving one row of counties northward.

As long as the total movement of isotherms toward the poles is much smaller than the size of the habitat, or the ranges in which the animals live, the effect on species is limited. But now the movement is inexorably toward the poles and totals more than one hundred miles over the past several decades. If greenhouse gases continue to increase at business-as-usual rates, then the rate of isotherm movement will double in this century to at least seventy miles per decade.

Species at the most immediate risk are those in polar climates and the biologically diverse slopes of alpine regions. Polar animals, in effect, will be pushed off the planet. Alpine species will be pushed toward higher altitudes, and toward smaller, rockier areas with thinner air; thus, in effect, they will also be pushed off the planet. A few such species, such as polar bears, no doubt will be "rescued" by human beings, but survival in zoos or managed animal reserves will be small consolation to bears or nature lovers.

Earth's history provides an invaluable perspective about what is possible. Fossils in the geologic record reveal that there have been five mass extinctions during the past five hundred million years—geologically brief periods in which about half or more of the species on Earth disappeared forever. In each case, life survived and new species developed over hundreds of thousands and millions of years. All these mass extinctions were associated with large and relatively rapid changes of atmospheric composition and climate. In the most extreme extinction, the "end-Permian" event, dividing the Permian and Triassic periods 251 million years ago, nearly all life on Earth—more than 90 percent of terrestrial and marine species—was exterminated.

None of the extinction events is understood in full. Research is active, as increasingly powerful methods of "reading the rocks" are being developed. Yet enough is now known to provide an invaluable perspective for what is already being called the sixth mass extinction, the human-caused destruction of species. Knowledge of past extinction events can inform us about potential paths for the future and perhaps help guide our actions, as our single powerful species threatens all others, and our own.

We do not know how many animal, plant, insect, and microbe species exist today. Nor do we know the rate we are driving species to extinction. About two million species—half of them being insects, including butterflies—have been cataloged, but more are discovered every day. The order of magnitude for the total is perhaps ten million. Some biologists estimate that when all the microbes, fungi, and parasites are counted, there may be one hundred million species.

Bird species are documented better than most. Everybody has heard of the dodo, the passenger pigeon, the ivory-billed woodpecker—all are gone—and the whooping crane, which, so far, we have just barely "saved." We are still losing one or two bird species per year. In total about 1 percent of bird species have disappeared over the past several centuries. If the loss of birds is representative of other species, several thousand species are becoming extinct each year.

The current extinction rate is at least one hundred times greater than the average natural rate. So the concern that humans may have initiated the sixth mass extinction is easy to understand. However, the outcome is still very much up in the air, and human-made climate change is likely to be the determining factor. I will argue that if we continue on a business-as-usual path, with a global warming of several degrees Celsius, then we will drive a large fraction of species, conceivably all species, to extinction. On the other hand, just as in the case of ice sheet stability, if we bring atmospheric composition under control in the near future, it is still possible to keep human-caused extinctions to a moderate level.

It's important to describe the specific extinction events through-out Earth's history, but first we should look at how information is gleaned from fossil records. Fossils are remains, impressions, or traces of life preserved in rock or sedimentary layers. All or portions of deceased organisms, from microscopic to dinosaur-size, and com-monly deposited on the floor of the ocean, lakes, bogs, and the allu-via of streams and rivers, are preserved in sediments and rocks that were formed after these deposits were buried under sufficient pres-sure. A variety of methods is now available to date these deposits, which makes it possible to catalog the history of species at many lo-cations. Extinctions are defined by the times beyond which fossils of a given species are not found anywhere in the world.

Causes of the end-Permian extinction, when life nearly died, are still debated, but some facts are reasonably clear. The extinction event took place during a time of massive volcanic eruptions in Siberia that spread basalt lava over an area the size of Europe in a layer as much as two miles thick. The lava outflow occurred over a period of about a million years. By itself the lava outflow probably could not be responsible for the extinctions—there have been a few larger lava outflows in Earth's history without such extreme loss of life. One factor may have been noxious gases from the eruptions, including acid rain produced by the volcanic sulfur dioxide emis-sions, which placed a stress on late Permian life-forms.

However, the biggest stress on life may have been the strong global warming that occurred at that time, about 6 degrees Celsius at low latitudes and probably more at high latitudes. This warming was a puzzle to scientists for some time, because the size and slow pace of the basalt eruption should not have produced enough carbon dioxide to yield such a great increase in temperature. It was only with the help of carbon isotope studies of end-Permian sediments that a likely explanation for the large magnitude of this global warming emerged.

Carbon isotopes are extremely important to climate studies. The properties of an element depend mainly on the number of protons in its nucleus. Some elements exist in more than one form (differ-ent isotopes), depending on the number of neutrons in the nucleus. The number of neutrons has only small effects on the element's properties, but the small effects turn out to be very useful for cli-mate studies. The most common form of carbon, carbon-12, has six protons and six neutrons. About 99 percent of the carbon atoms in carbon dioxide is carbon-12, and about 1 percent is carbon-13, with seven neutrons in the nucleus.

Plants prefer carbon-12, the light carbon. In other words, as plants grow by taking carbon dioxide out of the air, they take in more of the carbon dioxide that has carbon-12 than would be expected from its proportion in the air. Thus sedimentary deposits derived from biological material, such as coal, have an unusually large proportion of light carbon.

One of the characteristics of rocks formed during end-Permian time was an even greater proportion of light carbon. That meant the atmosphere at the time had an excessive amount of light carbon. How could that be? One possibility was that a huge amount of coal "burned" during that period—it was exposed to the surface and oxidized. But it would have required burning almost all the coal on the planet to reach such light carbon levels, and how could the coal have been unearthed? It did not seem plausible.

The likely source of the light carbon came to light in the past decade or so, as scientists began to focus on methane ice, also called methane clathrates or methane hydrates. This is essentially "frozen" methane, with each methane molecule enclosed in a "cage," or crystal, of water ice. Large amounts of methane ice are found today in arctic tundra (frozen ground) and, especially, beneath sediments on the seafloor of the Arctic Ocean. The methane was produced by bacterial degradation of organic matter in a low-oxygen environment—in other words, the rotting and decay of buried plant and animal remains—which yields an even a higher concentration of carbon-12 than coal does. This is the only carbon repository on Earth with enough light carbon to plausibly explain the end-Permian data.

The mystery of all the light carbon in the air during the end-Permian extinction finally was solved, but the details remain unclear. The large Siberian lava flows may have caused the melting and release of methane ice from the Siberian tundra and the ocean floor; or global warming due to carbon dioxide from the Siberian basalt eruptions caused melting of the methane ice, which then amplified the global warming, leading to the great warming of 6 or more degrees Celsius.

There are still many questions and theories about exactly what happened during the end-Permian extinction and why it was so devastating. Some scientists believe that an asteroid collided with Earth at that time, perhaps helping to initiate the Siberian basalt eruptions or methane hydrate release or both. But no convincing evidence of an asteroid collision has been provided. Most geologists agree that methane hydrates played a role, probably an important role. There is simply no other known source for such a large amount of light carbon. It also seems likely that the large global warming was an important factor in this great crisis for life on the planet.

What is certain is the magnitude of the devastation. It took about 50 million years for life to again develop the diversity that it had prior to the event.

Other extinctions, albeit less devastating ones, took place more recently and can be studied in more detail. The famous end-Cretaceous extinction, which wiped out about half the species on the planet, including the dinosaurs, occurred 65 million years ago when an asteroid struck Earth, producing a crater about one hundred miles wide on the Yucatán peninsula in Mexico. The extinctions are believed to have been caused, at least in part, by a massive injection of gas and dust into the atmosphere. Aerosols in the stratosphere would have blocked sunlight for a few years, reducing photosynthesis and causing a temporary global cooling.

A slightly more recent extinction event, about 55 million years ago, deserves greater scrutiny, because it is the most relevant to ongoing human actions. The Paleocene-Eocene thermal maximum (PETM) is classified as a minor extinction event—almost half the deep ocean foraminifera (microscopic shelled animals) species disappeared, but there was little extinction of land plants and animals. The range of some flora expanded poleward by hundreds and even thousands of miles. And the diversity, dispersal, and body sizes of terrestrial mammals changed rapidly.

Global warming of about 5 to 9 degrees Celsius occurred in the PETM, almost as great as in the end-Permian and comparable to the warming that may occur in the next century or so if business-as-usual greenhouse gas emissions continue. But the fact that most terrestrial species survived the PETM does not mean we shouldn't be concerned about the effect of future global warming, for two major reasons. First, the PETM warming occurred over millennia, not in a century. That means the power in the human punch is an order of magnitude greater. Climatic zones are moving poleward ten times faster now than in the PETM. Second, humans are simultaneously causing other stresses on animals and plants, by overharvesting, deforesting, and simply taking over large parts of the planet.

We need to dig deeper, to understand the PETM better, before drawing conclusions. But the PETM cannot be reliably interpreted in isolation. It needs to be looked at in the broader context of Earth's climate history, which has much to teach us.

That's easier said than done. The paleoclimate literature is voluminous and arcane. My own minor contribution, an empirical evaluation of climate sensitivity by comparing the last ice age and the Holocene, only scratched the surface of paleoclimate data. But in early June 2007 I received phone calls from the media that gave me an added push to dig a little deeper into paleoclimate.

The calls requested my reaction to a statement made by NASA administrator Michael Griffin on National Public Radio. This was Griffin's response to a question about global warming:

> I am not sure that it is fair to say that it is a problem we must wrestle with. To assume that it is a problem is to assume that the state of the Earth's climate today is the optimal climate, the best climate that we could have or ever have had, and that we need to take steps to make sure that it doesn't change. First of all, I don't think it's within the power of human beings to assure that the climate does not change, as millions of years of history have shown. And second of all, I guess I would ask which human beings—where and when—are to be accorded the privilege of deciding that this particular climate that we have right here today, right now, is the best climate for all other human beings. I think that's a rather arrogant position for people to take.

My reaction included "almost fell off my chair," "incredibly ignorant," and an assertion that surely it was in the common good to preserve species, sea level, and the climate zones that existed during the period that civilization developed. On the June 6 *Colbert Report*, Stephen Colbert showed some of Griffin's comments, then my response, and said, "There should be an interesting holiday party at NASA this year."

Upon reflection, I realized that many well-educated people might draw conclusions similar to Griffin's. It is not easy to appreciate the implications of paleoclimate time scales—Griffin obviously did not. But his ignorance underlined a broader problem. Paleoclimate data actually reveal the opposite of what Griffin concluded. So why have we been unable to make that clear, especially the staggering implications for global energy policies?

It seemed to me that part of the difficulty may be our emphasis on glacial-interglacial climate fluctuations—the periodic waxing and waning of large ice sheets on North America and Eurasia. To be sure, precise data on glacial-interglacial climate change obtained from ice cores—including which hemisphere a change originates in and how one quantity leads or another lags—are invaluable for sorting out its mechanisms and dynamics. But we researchers, as well as the public, might be able to see the forest for the trees better if we look at how the glacial-interglacial climate swings fit into longer-term, larger planetary change.

Such a longer time-scale perspective is provided by ocean cores. Ice sheets also existed on the planet millions of years ago, but they have long since melted, destroying their treasure of information. In contrast, for many millions of years there has been a slow rain of material sinking to the ocean floor, piling up in sediments. The most useful material in the sediments, for a climatologist, is the shells of the microscopic animals called foraminifera, or, for short, forams. The most useful characteristic of forams is their proportion of oxygen isotopes.

Hold on! This is not difficult! You already know that elements have different isotopes, depending on how many neutrons are in the nucleus. Almost 99.8 percent of oxygen is the garden-variety oxygen-16, with eight protons and eight neutrons. But about two tenths of 1 percent of oxygen is heavy oxygen-18, with 10 neutrons.

The great thing about oxygen-18 is that it gives us a thermometer, which we can use to measure Earth's temperature over hundreds of millions of years. All we have to do is measure the proportion of oxygen-18 in the dead bodies of critters that lived in the past.

What is wonderful about research these days (Benjamin Franklin would be enormously envious) is that we can store an enormous amount of data virtually on a pinhead, and we can transmit the data around the world in a second by using the Internet. In July 2007, when I decided that I would like to study oxygen-18 data for the

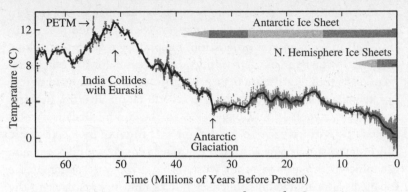

FIGURE 18. *Deep ocean temperature during the Cenozoic era.
(See text. Original data from Zachos et al., "Trends, Rhythms, and
Aberrations in Global Climate 65 Ma to Present." See sources.)*

Cenozoic era (that's the past 65 million years, from the time the
dinosaurs went extinct until today), I sent an e-mail to Jim Zachos,
perhaps the most prolific expert on Cenozoic climate. Within a few
days I received a remarkable data set for oxygen-18 covering the Cenozoic. Makiko Sato and I produced the final form of figure 18, with
oxygen-18 converted to temperature for the entire era, but Zachos deserves credit for the data and figure format. I do need to mention one
trick, or approximation, that we employed, so other scientists do not
beat me about the head and shoulders. First I must describe the data.

Ocean sediment cores have been extracted from many different
places all around the world. A core is obtained by pushing a very
long hollow pipe into the ocean sediments, capping the pipe, and
pulling it out. The sediments extracted were deposited at times extending from today (the top of the core) to millions of years ago, at
the bottom of a long core. The specific data set that I used was obtained from analyses on the cores' forams, the microscopically
small shelled critters living near the ocean floor.

Shells of forams are made of calcium carbonate $(CaCO_3)$. This
tiny critter grows its shell by taking calcium and carbon dioxide
from the water and snitching one oxygen atom from a water molecule. Water molecules in the ocean are all bouncing around, bang-
ing against each other, at a speed that depends on the temperature
of the water. The light water molecules, those with oxygen-16, are
moving faster than the heavier ones, and so they get incorporated

into the shell more easily. If the water gets warmer, the oxygen-16 gains even more speed relative to oxygen-18 and is incorporated in the shell in even greater proportion. Laboratory experiments show us just how fast the oxygen-16 portion increases (or oxygen-18 decreases) as temperature increases. Bingo—we have a thermometer.

Except for one catch. There is a second factor altering the proportion of oxygen-16 and oxygen-18 in the foram. Because water molecules with oxygen-16 are lighter and moving faster, they are more successful at penetrating the surface tension of the ocean and escaping to the air—in other words, they evaporate faster. If the escaped water molecule condenses out as rain, it goes back to the ocean, so the proportion of light oxygen in the ocean remains unaffected. But if the water molecules become snow and build an ice sheet, that ice sheet will have little oxygen-18. As the ice sheet gets bigger and bigger, the proportion of oxygen-18 remaining in the ocean gets bigger and bigger. So the amount of oxygen-16 and oxygen-18 in a foram shell depends on the size of global ice sheets as well as the temperature of the ocean water. This ambiguity spoils the thermometer, causing consternation among paleoclimate scientists. I made a simple assumption to deal with this ambiguity. Geologic data show that from the beginning of the Cenozoic until 34 million years ago, there were no large ice sheets on Earth, so the foram thermometer works without any correction in the early Cenozoic. From the time just before Antarctica froze over until the recent ice ages, the total change of oxygen-18 was twice as large as it would have been due to only the known temperature change between these two end points. So my simple assumption was that throughout the 34-million-year period the variations of oxygen-18 should always be assigned equally to temperature and ice volume change.

Okay, I will not bore you further. You can find details in our paper "Target Atmospheric CO_2: Where Should Humanity Aim?" freely available in *Open Atmospheric Sciences Journal* (2008). In that paper we showed, from independent sea level data, that apportioning changes of oxygen-18 equally between temperature and ice volume (sea level) worked well at both times when it could be checked: the rapid change when Antarctica froze over and the glacial-interglacial oscillations of the past several hundred thousand years. These are times when the assumption was most dubious, suggesting that it is a reasonable approximation for the full period.

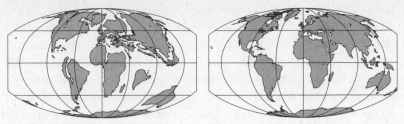

End of Cretaceous (65 MYBP)   Present Day

FIGURE 19. *Continental locations 65 million years ago and today. The Cretaceous era ended and the Cenozoic began 65 million years ago. (Data from Hansen et al., "Target Atmospheric CO_2," (see sources) based on original data from Ron Blakey at Northern Arizona University.)*

Even the most hardened antiscience zealot, once he understands figure 18, will have to admit that it is one of the most beautiful curves on the planet (I'm referring to scientific curves). It contains an enormous amount of interesting information about Earth's history. There are remarkable stories in both the broad sweep of climate over the 65 million years and in the rapid climate fluctuations.

First, note that the temperature increased in the early Cenozoic, reaching 13 degrees Celsius (55 degrees Fahrenheit) 50 million years ago. Then, over the last 50 million years, the planet cooled. In the past few million years, the coldest period in the record, glacial-to-interglacial oscillations became larger and larger. These temperatures were "measured" in the deep ocean, by the forams, but they tell us about the surface. The temperature in the deep ocean is the same as the temperature of the high-latitude ocean surface in the winter, because that is the season and place where ocean surface water becomes most dense and sinks to the ocean bottom.

Thirteen degrees Celsius in the winter at polar latitudes! Yes, Earth was much warmer 50 million years ago. Alaska had tropical-like vegetation and was occupied by crocodiles. Compare that with the recent ice ages—in some of them an ice sheet covered Canada and reached as far south as Kansas. As Administrator Griffin would point out, these are huge climate changes, and humans had nothing to do with them. *Homo sapiens* did not exist until the last two or three up-and-down blips at the right end of the figure 18 graph.

There are many stories in figure 18. There is information in the broad sweep of the curve, but also in the rapid climate oscillations.

Let us first consider the broad sweep, the great warming that peaked 50 million years ago, followed by a long cooling trend. What could have caused such a huge change of Earth's surface temperature? There are three possibilities: changes of the energy coming into the planet, changes on the surface, and changes in the atmosphere.

First, consider the energy coming in. Astronomers know that our sun is a very normal "main sequence" star. That phrase refers to a diagram that shows how a star changes as it ages. Our sun is a relatively young star, about 4.6 billion years old. It is still in the phase of "burning" hydrogen in its core, by nuclear fusion, making helium. In this phase the sun is slowly getting brighter. Over the past 65 million years, the sun's brightness has increased 0.4 percent. Earth absorbs about 240 watts (per square meter, averaged over the planet) of solar energy, so the solar forcing over the Cenozoic era has been a linear increase of about 1 watt. By itself, that should have caused a slow warming, of the order of 1 degree over 65 million years. But the planet has actually cooled, so the sun is not the biggest contributor to the climate changes in figure 18.

Second, consider how Earth's surface changed over the 65 million years. We know how continents were moving, in part from the orientation of Earth's magnetic field, as it was "frozen" into magnetized rocks that congealed at different times and places. Figure 19 compares the continental configuration at the beginning of the Cenozoic era and today. The Americas were closer to Europe and Africa in the early Cenozoic, and sea level was higher because of the absence of ice sheets, but the continents were close to their present latitudes. The location of continents affects the climate, mainly because the reflectivity of land is different than that of the ocean, which is very dark. However, the climate forcing due to changes in the arrangement of continents during the Cenozoic era is only of the order of 1 watt averaged over the planet.

Third, consider the changes in the atmosphere. The amount of atmospheric carbon dioxide during the Cenozoic varied from as little as 170 ppm in recent ice ages to 1,000 to 2,000 ppm in the early Cenozoic. Thus the largest carbon dioxide amount was probably close to three doublings of the smallest amount (170→ 340→ 680→ 1,360). Large carbon dioxide change is usefully expressed as the number of doublings, because the infrared absorption bands (illustrated in figure 5 on page 62) become saturated as carbon dioxide increases. Additional absorption occurs in weak bands and at the edges of strong absorption bands, but it takes more and more carbon dioxide to yield a given increment of climate forcing. The result is that forcing increases by about 4 watts with each doubling.

So carbon dioxide changes in the Cenozoic caused a forcing of about 12 watts—at least ten times greater than the climate forcing due to either the sun or Earth's surface. It follows that changing carbon dioxide is the immediate cause of the large climate swings over the last 65 million years.

Before we consider the reasons for this carbon dioxide change, it is important to check whether this greenhouse gas climate forcing is the correct order of magnitude to account for the measured change of Earth's temperature. If the topic of climate sensitivity is too esoteric for your taste, just skip the following three paragraphs. However, if you digest this stuff, it will help you understand the important matters in global climate change that the "professionals" are contemplating now.

Earth's temperature changed about 14 degrees Celsius between 50 million years ago and the recent ice ages (figure 18). Between 50 and 34 million years ago, the period when there were no large ice sheets on Earth, we expect climate sensitivity to be 3 degrees Celsius for doubled carbon dioxide (the empirical climate sensitivity we inferred earlier from glacial-interglacial climate change). That means a forcing of four thirds (4/3) of a watt is needed to cause a 1-degree Celsius temperature change. Thus the 8-degree temperature change between 50 and 34 million years ago required almost 11 watts of forcing. Between 34 million years ago and the depth of the last ice age, surface reflectivity change due to ice sheets approximately doubled the climate sensitivity (as discussed in chapter 3). Thus the 6-degree temperature change in that period required a greenhouse gas forcing of 4 watts. The greenhouse forcing required for the total temperature change over the Cenozoic is about 15 watts, assuming that climate sensitivity averages 3 degrees Celsius for doubled carbon dioxide in the absence of ice sheets.

Thus the estimated carbon dioxide forcing of 12 watts is, by itself, close to what is needed to account for the measured temperature change. Other long-lived greenhouse gases, specifically methane and nitrous oxide, are expected to augment the carbon dioxide forcing, that is, their atmospheric amount is likely to be greater when the planet is warmer. In addition, this is a good time to remind ourselves that the climate sensitivity of 3 degrees Celsius for doubled carbon dioxide was derived empirically from climate change in the late Cenozoic. At the warmest temperatures of the early Cenozoic it is likely that climate sensitivity was moving into a different, vitally important, climate regime, with higher climate sensitivity, as I will discuss in conjunction with data from the Paleocene-Eocene thermal maximum.

We must also note that the deep ocean temperature change defined by forams is not the same as global mean surface temperature change. The difference between the two must become large as the deep ocean temperature approaches the freezing point, because the deep ocean temperature does not go below the freezing point, while the surface continues to cool. However, even during the coldest increment in the entire Cenozoic curve (figure 18)—the time between the current interglacial period and the last ice age, when global surface temperature changed 5 degrees—there was a substantial deep ocean temperature change (3 degrees). But while global temperature change exceeded deep ocean temperature change in the late Cenozoic, the opposite is likely during the warm portion of the Cenozoic temperature curve. Why? Because high-latitude surface temperature change (which determines deep ocean temperature change) exceeds global mean temperature change—and ocean temperature was well away from the freezing point limitation in the early Cenozoic, so the amplified high-latitude temperature change was transmitted to the deep ocean. Thus overall, although there is necessarily uncertainty in the relation of global deep ocean temperature change and global surface temperature change, it appears that the total temperature changes over the Cenozoic era at the surface and in the deep ocean are comparable in magnitude.

Now let us turn to the question of why atmospheric carbon dioxide changed during the past 65 million years. First, note that the carbon dioxide causing the large climate changes in the Cenozoic era necessarily came from the solid Earth reservoirs (rocks or fossil fuels; see figure 15 on page 118). The alternative—oscillation of carbon among its surface reservoirs—is important for glacial-interglacial climate change, as a climate feedback, but it alters atmospheric carbon dioxide by only about 100 ppm, not 1,000 ppm.

The solid Earth is both a source of carbon dioxide for the surface reservoirs and a sink. The carbon dioxide source occurs at the edge of moving continental plates that "subduct" ocean crust. What does that mean? Continents are composed of relatively light material, typically granite. Ocean crust, that is, the solid Earth beneath the ocean water, is heavier rock, typically basalt. Both continents and ocean crust are lighter than material at greater depths, and they are slightly mobile because of convection deeper in the Earth. The energy that drives movement of the surface crust comes from the small amount of heat released by radioactive elements in Earth's interior. As continents move, commonly at a rate of several centimeters per year (an inch or two), they can ride over ocean crust. Intense heat and pressure due to the overriding continent cause melting and metamorphism of the ocean crust, producing carbon dioxide and methane from calcium carbonate and organic sediments on the ocean floor. The gases come to the surface in volcanic eruptions and at seltzer springs and gas vents. This is the main source of carbon dioxide from the solid Earth to surface reservoirs.

The main carbon sink—that is, the return flow of carbon to the solid Earth—occurs via the weathering of rocks. Chemical reactions combine carbon dioxide and minerals, with the ingredients being carried by streams and rivers to the ocean and precipitated to the ocean floor as carbonate sediments. A smaller, but still important, carbon sink is the sedimentation of organic material in the ocean, lakes, and bogs. Some of this organic material eventually forms fossil fuels and methane hydrates.

A key point is that the solid Earth source and the solid Earth sink of carbon are not in general equal at a given time. The imbalance causes the atmospheric carbon dioxide amount to vary. The carbon dioxide source to the atmosphere is larger, for example, when continental drift is occurring over a region of carbon-rich ocean crust.

A qualitative explanation for the large Cenozoic climate change, and a picture of the solid Earth's role in the Cenozoic carbon cycle, almost leaps out from figures 18 and 19. During the period between 60 and 50 million years ago, India was moving about 20 centimeters (8 inches) per year, which is unusually rapid for continental drift. India was heading north through an ocean region, now called the Indian Ocean, that had long been an area into which major rivers of the world had deposited carbon sediments. Undoubtedly, atmospheric carbon dioxide increased rapidly during that period as the carbon-rich sediments on that ocean floor were subducted beneath the Indian continental plate. Then, 50 million years ago, India crashed into Asia, with the Indian plate sliding under the Asian plate. The colliding continental plates began to push up the Himalayan mountains and Tibetan plateau, exposing a large amount of fresh rock for weathering. With India's sojourn across the carbon-rich ocean completed, the carbon dioxide emissions declined and the planet began a long-term cooling trend.

A quantitative analysis of the Cenozoic atmospheric carbon dioxide history is carried out in our "Target CO_2" paper described above. We calculated the range of carbon dioxide histories that can match the observed temperature curve (figure 18), accounting for uncertainties in the relation between the deep ocean and surface temperature. We estimated maximum carbon dioxide 50 million years ago as 1,400 ppm, with an uncertainty of about 500 ppm. The carbon dioxide amount 34 million years ago, when Antarctica became cold enough to harbor a large ice sheet, was found to be 450 ppm with an uncertainty of 100 ppm. This calculated carbon dioxide history falls within the broad range of estimates based on several indirect ways of measuring past carbon dioxide levels, as described in the "Target" CO_2 paper.

A striking conclusion from this analysis is the value of carbon dioxide—only 450 ppm, with estimated uncertainty of 100 ppm—at which the transition occurs from no large ice sheet to a glaciated Antarctica. This has a clear, strong implication for what constitutes a dangerous level of atmospheric carbon dioxide. If humanity burns most of the fossil fuels, doubling or tripling the preindustrial carbon dioxide level, Earth will surely head toward the ice-free condition, with sea level 75 meters (250 feet) higher than today. It is difficult to say how long it will take for the melting to be complete, but once ice sheet disintegration gets well under way, it will be impossible to stop.

With carbon dioxide the dominant climate forcing, as it is today, it obviously would be exceedingly foolish and dangerous to allow carbon dioxide to approach 450 ppm.

What does the Cenozoic history tell us with regard to Administrator Griffin's assertion that natural climate changes exceed human-made change?

Surely, nature changes carbon dioxide, and climate, by huge amounts. But we must look at time scales. The source of carbon dioxide emissions from the solid Earth to the surface reservoirs, when divided among the surface reservoirs, is a few ten thousandths of 1 ppm per year. The natural sink, weathering, has a similar magnitude. The natural source and sink can be out of balance, as when India was cruising through the Indian Ocean, by typically one ten thousandth of 1 ppm per year. In a million years such an imbalance changes atmospheric carbon dioxide by 100 ppm, a huge change.

But humans, by burning fossil fuels, are now increasing atmospheric carbon dioxide by 2 ppm per year. In other words, the human climate forcing is four orders of magnitude—ten thousand times—more powerful than the natural forcing. Humans are now in control of future climate, although I use the phrase "in control" loosely here.

Okay, I know, this is getting long, but for the sake of your children and grandchildren, let's look a little more closely at another story in figure 18, one that is vitally important. I refer to the PETM, the Paleocene-Eocene thermal maximum, the rapid warming of at least 5 degrees Celsius that occurred about 55 million years ago and caused a minor rash of extinctions, mainly of marine species.

The PETM looks like an explosion in figure 18, and by paleoclimate standards it was explosively rapid. Carbon isotopes in the sediments deposited during the PETM show that there was a huge injection of light carbon into the atmosphere—about 3,000 gigatons of carbon, almost as much as the carbon in all of today's oil, gas, and coal. It was injected in two bursts, each no more than a thousand years in duration.

The most likely source for such a rapid injection is methane hydrates. There is more than enough methane ice on continental shelves today to provide this amount of light carbon. The methane hydrate explanation is now broadly accepted, but it leaves open a vital question: What instigated the release of this methane? Was it an "external" trigger or a climate feedback? The answer holds enormous consequences for the future of humanity.

If the trigger for the methane hydrate release was external, such as the intrusion of hot magma from below or an asteroid crashing into the Arctic Ocean, then humans have no influence on whether the process will happen again. And the chances are remote that another such external event would happen in a time frame that most humans would care about. There have been several PETM-like rapid warming events in the past 200 million years. At that frequency, the chance of one beginning in the next hundred years is less than 0.00001 percent.

On the other hand, if the PETM and PETM-like methane hydrate releases were feedbacks, that is, if a warming climate caused the melting of frozen methane, then it is a whole different ball game. In that case, it is practically a dead certainty that business-as-usual exploitation of all fossil fuels would cause today's frozen methane to melt—it is only a question of how soon.

Unfortunately, paleoclimate data now unambiguously point to the methane releases being a feedback. If the PETM were an isolated case, that interpretation would be less certain. But it has been found that several PETM-like events in the Jurassic and Paleocene eras were, as with the PETM, "astronomically paced." Huh? That means the spikes in global warming and light-carbon sediments occurred simultaneously with the warm phase of climate oscillations caused by perturbations of Earth's orbit. In other words, the methane releases occurred at times of natural warming events

So, why do methane hydrates produce a huge amplifying feedback in a small number of cases, while most "astronomical" warmings show little or no evidence of methane hydrate amplification? That mercurial behavior, in fact, is exactly what is expected for methane hydrates.

The largest volume of methane hydrates is on continental shelves, in the top several hundred meters of ocean sediments, although a smaller volume exists in continental tundra. The marine methane hydrates form in coastal zones with high biologic productivity. A sufficient rain of organic material onto the ocean floor yields a low-oxygen environment in the sediments, which causes the bacterial degradation of organic matter to produce methane. If the temperature is right, the methane is frozen into hydrates.

If a warming occurs that is large enough to melt methane hydrate, each liter of melted hydrate expands into 160 liters of methane gas.

A small methane release may dissolve in the ocean, but a large release can bubble to the surface. Methane is a strong greenhouse gas, and on a time scale of about a decade it is oxidized to carbon dioxide, which will continue to cause warming for centuries. If the warming is large enough, most of the methane hydrate on continental shelves may be melted, as seems to have been the case in the PETM.

If Earth's methane hydrate inventory is suddenly discharged, as during the PETM event, it requires several million years to fully reload the planet's methane hydrate gun. Thus the next light-carbon methane hydrate event in the Paleocene, about 2 million years after the PETM, was only about half the strength of the PETM. This half-PETM was followed by still weaker and more frequent light-carbon warming spikes. These events occurred in conjunction with astronomical warming peaks during the time Earth was on its track toward peak warmth 50 million years ago, which suggests that the warmer Earth made the melting of hydrates easier and did not allow the hydrate reservoir to return to pre-PETM size.

Today, following global cooling over tens of millions of years, the methane hydrate reservoir is fully charged. The size of the hydrate reservoir is difficult to determine from spotty field data. However, methane hydrate models that are consistent with the limited data suggest a total inventory of about 5,000 gigatons of carbon in the form of methane ice and methane bubbles. Thus, unfortunately, not only is the methane gun now fully loaded, but it also has a charge larger than the one that existed prior to the PETM blast.

Let's not jump to conclusions, however. We must glean more from the PETM before we discuss the likely fate of today's frozen methane. Comparisons of the timing of carbon and temperature changes at many ocean sites show that a dramatic change in ocean circulation occurred at the time of the rapid PETM increases of light carbon and temperature. The ocean circulation change indicates that the main location where dense surface water sank toward the ocean bottom shifted from the region around Antarctica to middle latitudes in the northern hemisphere. Sinking water at the new location was also dense, but warmer and saltier. It is likely that this warmer water instigated the melting of methane hydrates. The methane, and carbon dioxide that formed as methane oxidized, provided an amplifying feedback that resulted in the large PETM spike in global temperature.

Why ocean circulation changed is uncertain, but it is likely related to the global warming of 2 to 3 degrees Celsius that occurred just prior to the PETM event (figure 18).

One final mundane, but sobering, inference from the PETM: The recovery time from excess carbon in the air and ocean, and from the PETM global warming spike, was about 100,000 years. That is the recovery time predicted by carbon cycle models. The added carbon dioxide in the air increases the rate of weathering and carbon uptake, which is a negative (diminishing) feedback. Confirmation of the recovery time is a useful verification of the models. It is also a reminder that if humans are so foolish as to burn all fossil fuels, the planet will not recover on any time scale that humans can imagine.

Such was the state of PETM research, or at least my perspective on it, in mid-2007, right around the time that Bill McKibben was asking me about 450 ppm—though the most startling revelation from the PETM was yet to come. I finally promised Bill that I would give him a number at the December 2007 American Geophysical Union meeting, when I would present a talk on the rationale for the suggested carbon dioxide target.

In that talk, I emphasized carbon dioxide itself, not the carbon dioxide equivalent of all human-made gases. The perturbed carbon cycle will not recover for tens of thousands of years, and it is carbon dioxide that determines the magnitude of the perturbation. Other forcings are important and need to be minimized, and some may be easier than carbon dioxide to deal with, but policy makers must understand that they cannot avoid constraints on carbon dioxide via offsets from other constituents.

In addition to paleoclimate data, my talk covered ongoing observations of five phenomena, all of which imply that an appropriate initial target should be no higher than 350 ppm. In brief, here are the five observations.

(1) The area of Arctic sea ice has been declining faster than models predicted. The end-of-summer sea ice area was 40 percent less in 2007 than in the late 1970s when accurate satellite measurements began. Continued growth of atmospheric carbon dioxide surely will result in an ice-free end-of-summer Arctic within several decades, with detrimental effects on wildlife and indigenous people. It is difficult to imagine how the Greenland ice sheet could survive if Arctic sea ice is lost entirely in the warm season. Retention of

warm-season sea ice likely requires restoration of the planet's energy balance. At present our best estimate is that there is about 0.5 watt per square meter more energy coming into the planet than is being emitted to space as heat radiation. A reduction of carbon dioxide amount from the current 387 ppm to 350 ppm, all other things being unchanged, would increase outgoing radiation by 0.5 watt, restoring planetary energy balance.

(2) Mountain glaciers are disappearing all over the world. If business-as-usual greenhouse gas emissions continue, most of the glaciers will be gone within fifty years. Rivers originating in glacier regions provide fresh water for billions of people. If the glaciers disappear, there will be heavy snowmelt and floods in the spring, but many dry rivers in the late summer and fall. The melting of glaciers is proceeding rapidly at current atmospheric composition. Probably the best we can hope is that restoration of the planet's energy balance may halt glacier recession.

(3) The Greenland and West Antarctic ice sheets are each losing mass at more than 100 cubic kilometers per year, and sea level is rising at more than 3 centimeters per decade. Clearly the ice sheets are unstable with the present climate forcing. Ice shelves around Antarctica are melting rapidly. It is difficult to say how far carbon dioxide must be reduced to stabilize the ice sheets, but clearly 387 ppm is too much.

(4) Data show that subtropical regions have expanded poleward by 4 degrees of latitude on average. Such expansion is an expected effect of global warming, but the change has been faster than predicted. Dry regions have expanded in the southern United States, the Mediterranean, and Australia. Fire frequency and area in the western United States have increased by 300 percent over the past several decades. Lake Powell and Lake Mead are now only half full. Climate change is a major cause of these regional shifts, although forest management practices and increased usage of freshwater aggravate the resulting problems.

(5) Coral reefs, where a quarter of all marine biological species are located, are suffering from multiple stresses, with two of the most important stresses, ocean acidification and warming surface water, caused by increasing carbon dioxide. As carbon dioxide in the air increases, the ocean dissolves some of the carbon dioxide, becoming more acidic. This makes it more difficult for animals

with carbonate shells or skeletons to survive—indeed, sufficiently acidic water dissolves carbonates. Ongoing studies suggest that coral reefs would have a better chance of surviving modern stresses if carbon dioxide were reduced to less than 350 ppm.

I am often asked: If we want to maintain Holocene-like climate, why should the target carbon dioxide not be close to the preindustrial amount, say 300 ppm or 280 ppm? The reason, in part, is that there are other climate forcings besides carbon dioxide, and we do not expect those to return to preindustrial levels. There is no plan to remove all roadways, buildings, and other human-made effects on the planet's surface. Nor will we prevent all activities that produce aerosols. Until we know all forcings and understand their net effect, it is premature to be more specific than "less than 350 ppm," and it is unnecessary for policy purposes. It will take time to turn carbon dioxide around and for it to begin to approach 350 ppm. By then, if we have been making appropriate measurements, our knowledge should be much improved and we will have extensive empirical evidence on real-world changes. Also our best current estimate for the planet's mean energy imbalance over the past decade, thus averaged over the solar cycle, is about +0.5 watt per square meter. Reducing carbon dioxide to 350 ppm would increase emission to space 0.5 watt per square meter, restoring the planet's energy balance, to first approximation.

There is a longer story and range of uncertainty for each of the five phenomena discussed above. The way science works, we must expose the caveats and keep an open mind—otherwise we will not be successful in the long run. I know you do not want a long story, so I will provide a flavor, by an example. The example also shows how people who are determined to discredit the threat of human-made climate change—I call them contrarians; others call them denialists—use uncertainties inappropriately to cast doubt on all conclusions, even those that can be made with confidence. Nobody has figured out a good way to deal with this problem, but we cannot change the way we do science, so we just have to present the data as best we can.

Let's look at Arctic sea ice as an example. Figure 20 shows the area of sea ice remaining at the end of the warm season (September in the northern hemisphere). The fate of summer sea ice is important. Loss of the ice would affect the stability of the Greenland ice sheet, the stability of methane hydrates in the ocean sediments and

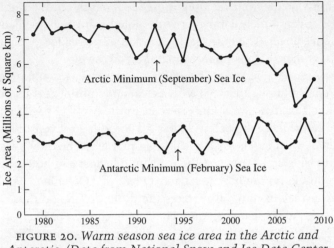

FIGURE 20. *Warm season sea ice area in the Arctic and Antarctic. (Data from National Snow and Ice Data Center Web site, http://nsidc.org/data/seaice_index/daily.html.)*

tundra, and species viability. Note in the graph that ice area fluctuates a lot from year to year—that's expected; the atmosphere and ocean have significant "weather noise," i.e., unforced and unpredictable chaotic variability.

Through 2006, Arctic sea ice was nearly following the script predicted by climate models. Sea ice area was beginning to decrease, just a bit faster than most models predicted. Then, in 2007, the bottom fell out. There was a big melt-off that surprised everyone. The ice area at the end of the warm season was only about 4 million square kilometers; three decades earlier, when accurate satellite measurements were initiated, it was 7 to 8 million square kilometers. Climate models had not predicted such a large loss before the middle of the twenty-first century.

A few (very few) scientists then suggested that summer sea ice might be gone entirely in five or six years. Those politicians who believe that scientists are inherently reticent, understating dangers, jumped on that speculation as if it were fact. But, as you can see in figure 20, the sea ice area partly recovered in 2008 and 2009. Contrarians, as is their wont, leaped on the recovery as evidence that there is no basis for concern. They also trumpeted that Antarctic sea ice is increasing rapidly. In fact, there are reasons to expect little change in Antarctic sea ice in the near term—the discharge of cold fresh water from disintegrating ice shelves tends to increase sea ice cover, which competes with global warming's tendency to reduce ice cover. The bottom line for Antarctic sea ice, as figure 20 shows, is that there is no meaningful trend as yet.

The Arctic is the issue. There is a strong consensus among Arctic researchers that we are faced with a clear and imminent threat to the continued existence of summer sea ice in the Arctic. I have found no Arctic researcher who believes that sea ice will survive if the world continues with business-as-usual fossil fuel use. The only questions seem to be exactly how fast the ice would be lost and how dramatic the feedbacks on tundra, methane hydrates, and Greenland would be.

The sea ice example illustrates the difficulty in communicating with the public. Contrarians spout their interpretations of data, sometimes mangling the truth, usually demonstrating a lack of insight about what is important, and often succeeding in confusing the public. Contrarians have a loud voice, out of proportion with their scientific standing, in part because of support from special interests and politicians influenced by special interests, and often aided by media, which likes to present two sides of every topic, creating the impression that the contrary opinions deserve equal respect. What can we expect the public to think when they compare a scientist who includes appropriate caveats with a contrarian who gives conclusions without hesitation? It can seem like a debate between theorists, and often the contrarians are more media savvy. It is no wonder that there is a growing gap between what is understood about global warming by the relevant scientific community and what is known about global warming by the public and policy makers.

What can be done to improve this situation? There is no simple good answer, or it would have been found by now. One suggestion that I have made repeatedly is that President Obama ask the National Academy of Sciences for a report on the status of climate science and its implications for policy makers. The academy, established by Abraham Lincoln for just such advisory purposes, is among the most respected scientific bodies in the world. Given the cacophony about global warming in the media, such authoritative guidance is needed to help define appropriate policies and to inform the public—but unless the report is specifically requested by the president, it will not have much impact.

Can scientists help improve communication so the public can better assess these matters? It is said that the public has lost interest in science, and that may be so. But we still have to try to communicate, using the same language, which requires a mutual effort. I hope that more of the public will be willing to look at straightforward scientific graphs of data. Graphs are the most compact, honest way of presenting information, allowing insights about what the data show and helping us distinguish what is significant and what is less important. They can help us assess where the climate itself is headed and how the driving factors are changing. Are human-caused climate forcings continuing on a business-as-usual course, or are they beginning to turn toward a path that can stabilize the climate? Is there evidence that amplifying feedbacks are moving toward runaway self-amplification, or are these feedbacks diminishing? Data are a work in progress because some of the most important quantities are not being measured, or are being measured with poor accuracy. Also, we are dealing with science on the fly—new quantities of importance may emerge, resulting in additional graphs. These and graphs included in this book will be updated regularly on my public Web site, with data sources provided, so the public can see how things are changing.

Global temperature must be one of the climate diagnostics, but it is a product of many driving factors and contains a good amount of variability that has nothing to do with climate forcings. By looking at the temperature data, we can avoid the common mistake of confusing local fluctuations with global climate change. For example, the summer of 2009 was unusually cool throughout much of the United States, which provided a field day for the contrarians in their efforts to confuse the public. Let's consider the data.

Figure 21 is a global map of surface temperature anomalies for June to August 2009 (summer in the northern hemisphere). The temperature anomaly is the difference between the actual June–August temperature in 2009 and the average June–August temperature between 1951 and 1980. That thirty-year period for climatology seemed appropriate as a point of reference when global warming first became an issue in the 1980s, and it seems best to continue using it as a fixed reference rather than have a continually shifting base period. Also, it makes the reference period the time when the post–World War II baby boomers grew up, a time that many of to-day's adults can remember.

Figure 21 shows that the region of low temperature in the United States and Canada was the exception, not the rule during the summer of 2009. Indeed, the global average anomaly for June–August 2009 was +0.6 degree Celsius, making it the second warmest in the period of instrumental data (1880–2009). It may have been a cool summer in the U.S., but unusual cool in one spot does not mean that global warming has gone away. People in the United States need to remember that the entire contiguous forty-eight states represent only 1.5 percent of Earth's surface.

The message is that we must not confuse weather and climate, which is the average weather. A three-month average, because three months is too brief to average out the effect of slow-moving weather systems, still contains a large amount of weather noise. That is the reason for the blobs of negative and positive temperature anomalies in figure 21. (My relatives in the Midwest just happened to be living under the coldest blob in the world, relative to normal local temperatures, during the summer of 2009.) As a result,

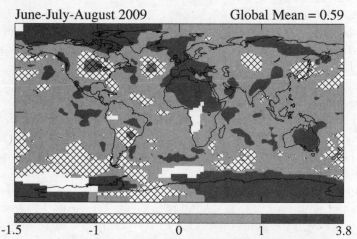

FIGURE 21. *Temperature anomalies in the 2009 northern hemisphere summer, relative to 1951–1980. It was the second warmest summer in 130 years, but the coldest anomaly fell over the United States. White areas are regions without observations. (Data update of Hansen et al., "GISS Analysis of Surface Temperature Change." See sources for chapter 6.)*

when I used a colored die to represent the effect of global warming on the frequency of warm seasons, I showed that one side of the die for the present decade is still blue—at any given location we expect a given season to have about a one-in-six chance of being unusually cool relative to the 1951–1980 climatology.

The gap between public perception and scientific reality is now enormous. While some of the public is just becoming aware of the existence of global warming, the relevant scientists—those who know what they are talking about—realize that the climate system is on the verge of tipping points. If the world does not make a dramatic shift in energy policies over the next few years, we may well pass the point of no return.

CHAPTER 9

An Honest, Effective Path

A SIMPLE, CLEAR, URGENT CONCLUSION leaped out from our research on the appropriate target level of atmospheric carbon dioxide: Coal emissions must be phased out as rapidly as possible or global climate disasters will be a dead certainty. The rationale for that statement was straightforward. But would it be clear to the people who need to know, the public and policy makers?

People were well aware of the global warming issue, thanks in no small part to Al Gore's 2006 movie *An Inconvenient Truth*. But even those fully persuaded about the reality of the climate threat did not seem to understand the principal implication. The public and policy makers concluded that they should slow down their rate of fossil fuel use, or at least the growth rate of that usage. For example, they should resolve to drive a more fuel-efficient vehicle, change their lightbulbs, add insulation, and so on. Or, if reducing personal emissions was inconvenient, they could purchase "offsets"—for example, they could pay other people to reduce *their* emissions. The planet would come out fine, right?

Wrong. The problem is that the act of slowing down emissions, by itself, does almost no good. The reason is that the lifetime of carbon dioxide added to the atmosphere-ocean system is millennia. So it does not matter much whether the fossil fuel is burned this year or next year. Energy efficiency is certainly an essential part of the solution to global warming, but it must be part of a strategic approach that leaves most of the fossil fuels in the ground.

Yes, most of the fossil fuels must be left in the ground. That is the explicit message that the science provides. Once science has delivered this conclusion, should the scientist leave it at that, allowing the politicians to deal with the problem? Any doubt about the right answer to that question should be erased by the experiences I will relate in this chapter.

The amount of carbon in the three conventional fossil fuels, oil, gas, and coal, is shown in figure 22. The black portions are the amount of fuel that has been burned already. Remaining reserves are uncertain and depend on whether we will go to Earth's extremes to get every last drop we can find. The estimates of the Intergovernmental Panel on Climate Change (IPCC) are probably representative of the readily available large pools of oil and gas. The larger estimates of the U.S. Energy Information Administration (EIA) may be more appropriate if fossil fuel companies are encouraged to get every last drop, by allowing them access to public lands, offshore areas, the deep ocean, and the Arctic, for example. The coal reserve estimate is from the World Energy Council. Although coal reserves are uncertain, we know there is plenty of coal to take the planet far into the dangerous zone, guaranteeing climate disasters.

Figure 22 also shows an estimate for the net amount of carbon that humans have added to the atmosphere from land use—with the primary effect being deforestation that is partially balanced by regrowth. The vertical whisker is an indication of the substantial uncertainty in the net land-use emissions. In the future the land-use bar may get bigger from further deforestation, or it could decrease with the help of improved forestry and agricultural practices.

Unconventional fossil fuels such as tar sands, shale oil, and methane hydrates are not included in figure 22. So far these fuels have barely been tapped and have contributed little to carbon in the air, but their estimated reserves are even greater than those of coal. Policy makers need to understand that these unconventional fossil fuels, which are as dirty and polluting as coal, must be left in the ground if we wish future generations to have a livable planet.

If coal emissions are phased out rapidly—a tall order, but a feasible one—the climate problem is solvable. It is coal emissions that must be eliminated, not necessarily coal use. If the carbon dioxide from coal burning can be captured and safely stored, coal, in principle, could continue to be used. But "carbon capture and sequestration," as it is called, makes coal use less efficient and much more expensive, because of the energy needed to capture and store the

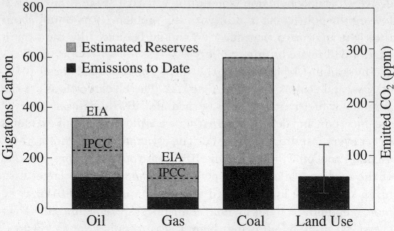

FIGURE 22. *Fossil fuel and net land-use emissions (1751–2008). (Data from Hansen et al., "Target Atmospheric CO$_2$." See sources for chapter 8.)*

carbon dioxide. It is also important to keep in mind that this process will not eliminate all air and water pollution from coal, nor will it eliminate damage due to coal mining. "Clean coal" is an oxymoron. The clean-coal concept, at least so far, has been an illusion, a diversion that the coal industry and its government supporters employ to allow dirty-coal uses to continue. Present efforts to develop carbon capture and storage, in Germany, the United States, and elsewhere, are more serious, but whether the technology will ever be accepted and adopted on a large scale—given its likely high cost, possible leakage of carbon dioxide from storage sites, and environmental damage from coal mining—are open questions.

There is no need to debate whether carbon capture and sequestration is realistic. The science demands a simple rule: Coal use must be prohibited unless and until the emissions can be captured and safely disposed of. If such a requirement were in place, it is uncertain whether utilities would build more coal-fired power plants—but that decision can be left to the marketplace. The point is that for the sake of our children and grandchildren, we cannot allow our government to continue to connive with the coal industry in subterfuges that allow dirty-coal use to continue.

Figure 23 provides quantitative verification of what is possible if

coal emissions are phased out rapidly. It shows expected future levels of atmospheric carbon dioxide under the assumption that world
coal emissions are phased out between 2010 and 2030 (linearly—
meaning that coal emissions would be cut in half by 2020). With
such a coal phaseout, carbon dioxide would reach a peak of "only"
400 to 425 parts per million sometime in the first half of this century.

Carbon dioxide would peak early (about 2025) and at around 400
ppm in the case of the IPCC oil and gas reserves estimate (figure
22). This lower estimate of reserves is relevant if we do not go after
every last drop in the ground but instead focus on developing energy sources for the era "beyond fossil fuels." In this optimistic
case, it would be possible to bring carbon dioxide back below the
350 ppm level by the end of this century via an extensive effort to
increase storage of carbon in forests and soils. An even earlier return to 350 ppm is conceivable via further actions such as the use
of carbon capture and storage at power plants that burn gas, oil, or
biofuels. (These cases are discussed in more detail in our 2008 "Target Atmospheric CO_2," paper, along with the appropriate qualifica-
tions and caveats, especially in the supplementary material.)

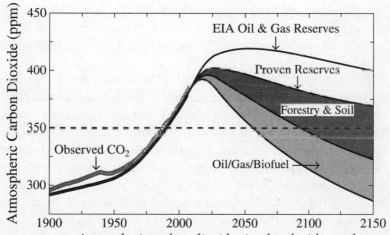

FIGURE 23. *Atmospheric carbon dioxide simulated with a carbon
cycle model under the assumption that coal emissions are phased
out over the period 2010–2030. Future carbon dioxide levels
depend on the size of oil and gas reserves and on other potential
actions. (Data from Hansen et al., "Target Atmospheric CO_2."
See sources for chapter 8.)*

Projections such as those in figure 23 are based on models of the carbon cycle that have various uncertainties, including the degree to which Earth's system will continue to be able to take up carbon when climate change accelerates. These uncertainties are important, but they should remain relatively small if climate change is kept to a minimum, as it would be in the coal phaseout scenarios that we investigate. I also want to emphasize that the use of biofuels should not be at the expense of food crops. Biofuels make global sense only when they are grown on marginal or degraded land or made with fuel derived from waste material.

Okay, we have shown that, by phasing out coal use, it is possible to keep maximum carbon dioxide close to 400 ppm, and in a period of several decades to get it back to 350 ppm and below. But why do we say that a coal phaseout is the *only* way to do it? Could we not instead stop using oil and gas immediately, while continuing to use coal (for a while)?

No. That is not plausible, and here's why: The large pools of oil and gas are owned by Russia and Middle East countries such as Saudi Arabia. How would we convince them to leave their oil in the ground? It is not going to happen. Besides, we would not want it to happen. We just barely have time to phase in technologies for the era beyond fossil fuels, even if we begin now with an "all hands on deck" strategy. We're simply not ready to suddenly stop using gas and oil.

So, if we want to solve the climate problem, we must phase out coal emissions. Period.

But is it feasible to phase out coal—does it make sense? Actually, it is *not* phasing out coal that makes no sense. Coal is exceedingly dirty stuff. Its mercury, arsenic, sulfates, and other constituents are a major source of global air and water pollution, leading to increased birth defects, impaired intelligence, asthma, and other respiratory and cardiovascular diseases. Coal's effect on air and water pollution is global—nobody escapes its reach. Mercury and other pollutants are deposited on land and in the ocean, infiltrating the food chain and building up in the bodies of long-lived animals and fish.

Coal's global pollution effects are compounded by the devastating regional effects of the various techniques for dredging the dirty stuff to the surface. The most barbaric approach, mountaintop removal, can only be described as blasphemous, whether or not na-

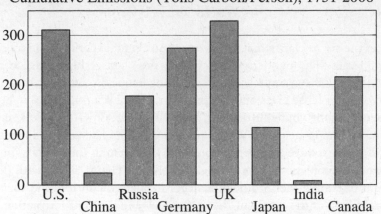

FIGURE 24. *Cumulative per capita carbon dioxide emissions, with countries listed in the order of national cumulative emissions. (Data sources are Carbon Dioxide Information Analysis Center, Oak Ridge National Laboratory, and British Petroleum.)*

ture is one's only religion. Mountaintop-removal mining does more than irreparably scar our mountain ranges. Toxic sludge ponds and mining waste dumped into valleys poison the water supply, causing multiple documented health problems for nearby populations.

While mountaintop removal is an emotional topic, the focus needs to be on the big picture. What policies are needed to rapidly phase out carbon dioxide emissions?

You will hear politicians and others say something like, "We have a plan. We will reduce emissions 25 percent by 2020, 90 percent by 2050." Or they will give some other numbers. But the numbers are meaningless, as you can easily prove. Just ask them this question: "Are you going to continue to use coal, and maybe even permit another coal plant to be built?" If they say yes, then ask them how they plan to convince Russia and Saudi Arabia to leave their oil in the ground. When they tell you that they are going to solve the problem via a "goal," "binding target," or a "cap," you know that they are lying. Yes, *lying* is a harsh word, so you may instead say "kidding themselves." But I expect that one day your more perceptive grandchildren will say that you let the politicians lie to you.

World Travels

Let me try to clarify matters by recounting my experiences in a few countries. The publicity surrounding the "censorship" episode of January 2006 generated many speaking requests, which I mostly avoided because I was busy working on several scientific papers. But one engaging young British entrepreneur, a certain George Polk, entreated me to help him with a problem that he had set for himself: educating a large number of potentially influential individuals, in a range of professions, about the global climate change issue, with the aim of getting them to help push for "concrete political and social steps to shift the excellent rhetoric we have in the U.K. into action."

That objective appealed to me. All countries have the same problem: Politicians talk about environmentalism, but their actions are inconsistent with the talk. The U.K. seemed to be the right place to press for real action. The prime minister at the time, Tony Blair, often spoke of the need to combat global warming. And if the U.K. were persuaded about the need for specific actions, it conceivably might have some influence on the United States, its long-standing ally.

I had one additional argument for giving a speech in the U.K.: On a per capita basis, the U.K. is more responsible for the climate problem than any other nation. That may be surprising, given that the U.K. produces less than 2 percent of global fossil fuel emissions today—the United States and China each burn more than ten times as much fossil fuels. But climate change is caused by cumulative historical emissions. The fraction of carbon dioxide emissions remaining in the air today is much less for older emissions than for recent emissions, due to carbon uptake by the ocean and biosphere. But the greater diminishment of older emissions is compensated by the fact that they have had more time to affect climate. The result is that the U.K., United States, and Germany, in that order, are the three countries most responsible, per capita, for cumulative emissions and climate change, as shown quantitatively in figure 24.

I gave talks in London in March and July of 2007. Both trips included dinners with people who, it was hoped, might be able to influence policies. At one of the dinners, on U.S. Independence Day (July 4), my argument that new coal-fired power plants must be stopped as a first step toward phasing out coal emissions evoked discussion in favor of a specific action. The group would write a letter to try to influence plans to expand an existing coal-fired power plant. I contributed science rationale for the letter.

Later in the year, as it was clear that plans for expansion of coal-fired power were continuing apace in the U.K., Germany, and the United States, and even more so in China and India, I decided to write a letter to the U.K. prime minister, by then Gordon Brown. If Prime Minister Brown wanted to exert leadership in the climate problem, I wrote, a moratorium on new coal-fired power would be the way to do it, and would be more effective than a "goal" for emissions reduction. Such an action would put the U.K. in position to argue that Germany and the United States, both planning to build more coal plants, should also have a moratorium. Until Europe and the United States stop building new coal plants, there is little chance of fruitful discussions with China or India—and no hope of solving the climate problem.

The letter was taken by U.K. contacts to appropriate people in the prime minister's office. My hope was to get the U.K. government to think about the problem in a different way, to recognize that goals for emission reduction, however ambitious, will not work. There is a limit on how much carbon dioxide we can put into the air, and the only realistic chance of staying under that limit is to cut off coal emissions to the atmosphere soon. I believe my letter to the prime minister, which I made public on my Web site, was the clearest explanation that I had made of this concept. Unfortunately, the official response, via a letter to me from the Department of Environment, was tantamount to a restatement of the U.K.'s prior positions and plans.

I had already decided to write a letter to German chancellor Angela Merkel, before receiving a response from the U.K. government. Merkel was trained as a physicist, and I hoped that, rather than relying on advisers, she would be willing to think about the problem herself. I figured she would be able to appreciate the geophysical boundary conditions, the conclusion that most of the coal must be left in the ground.

My letter to Chancellor Merkel was similar to my letter to Prime Minister Brown. I sent a draft of the letter to German scientists and environmentalists, who provided helpful suggestions on details and protocol to optimize the chances that my suggestion would be considered seriously. A German environmental organization, GermanWatch, arranged translation of the letter into German for publication in a major German newspaper, *Die Zeit*.

An open letter, it seemed to me, was probably the best way to affect the coal discussion in Germany. However, John Schellnhuber, climate science adviser to the German government, suggested that it would be better if I held off on publishing the German translation and instead traveled to Germany for discussions. Specifically, Schellnhuber argued that the only way to get Germany to change its position was to persuade Sigmar Gabriel, minister for the environment, of the need to do that. Gabriel was in charge of German efforts to control the country's greenhouse gas emissions, and Merkel relied on him for policy advice.

My trip to Germany began with a useful visit to the Potsdam Institute for Climate Impact Research, of which Schellnhuber is director. I gave a seminar at the institute to initiate a discussion, the objective being to make sure that we were in basic agreement on the science issues to be addressed with Minister Gabriel. The scientists at the Potsdam Institute include Stefan Rahmstorf, one of the world's leading researchers on climate change. Rahmstorf has a broad understanding of the science and an ability to explain things to the nonscientist, so it was useful that he agreed to attend the meeting with Gabriel.

That meeting lasted about ninety minutes. In the first part of the discussion I provided reasons for a carbon dioxide target of no more than 350 ppm—explaining the need to avoid ice sheet disintegration, species extinction, loss of mountain glaciers and freshwater supplies, expansion of the subtropics, increasingly extreme forest fires and floods, and destruction of the great biodiversity of coral reefs. This part of the meeting went very well. There was no disagreement about the need to aim for a limit of 350 ppm on carbon dioxide.

The sticking point was the implication: the need to halt coal emissions. Germany had plans to build several more coal-fired power plants. Our discussion circled back to this issue repeatedly, as I argued that the climate problem could not be solved if coal use continued. I asked how, if Germany was going to continue to burn coal, it would persuade countries such as Russia to leave its oil in the ground. Gabriel's answer: "We will tighten the carbon cap." I pointed out that a cap slows emissions but does not prevent usage of the large pools of oil and gas reserves. A cap, or a carbon tax, is useful—indeed, necessary—to spur technologies needed to supplant fossil fuels, so that marginal reserves (fossil fuels that are difficult to extract) can be left in the ground. However, a cap does not prevent readily accessible oil and gas from being extracted. We went around this circle several times.

Then, in the final minutes of our meeting, the underlying story emerged with clarity: Coal use was essential, Minister Gabriel said, because Germany was going to phase out nuclear power. Period. It was a political decision, and it was not negotiable.

As we stood up to leave, Gabriel asked me whether I had an appointment to see Merkel. He seemed satisfied that I did not. On the trip back to the United States I had the feeling that perhaps I had missed an opportunity. I should have been more involved in defining the arrangements for the trip—conceivably I could have obtained a meeting with Merkel, especially if I had made such a meeting a condition for withdrawing the letter from the media. That leverage had been lost by the time of my meeting with Gabriel, as the letter was old news by then.

Two weeks later I went to Japan, where I had been asked to give a keynote talk at a symposium at the United Nations University, whose main campus is in Tokyo. The symposium's purpose, as described in the letter requesting my talk, was "to raise public awareness in Japan and internationally about the challenges and emerging approaches to climate change in advance of the forthcoming G8 Summit to be held in Hokkaido, Japan. The specific objective is to bring together some of the world's leading scientists and writers on environmental issues, groundbreaking thinkers at the intersection of science and communications, to examine how our thinking needs to change if we are to collectively take on the myriad challenges presented by global warming."

That was an opportunity—or an obligation—that I should not turn down. Japan was the perfect place to clarify the implications of science for policy. It was the birthplace of the Kyoto Protocol, the first attempt at international climate policy, which expires in 2012. Before a follow-up international agreement is defined, the scientific method and common sense suggest that we look at evidence from that first attempt.

So I would focus my lecture at the UN University on lessons from Kyoto. I also would write a letter to Japanese prime minister Yasuo Fukuda. Japan—which had been exemplary in living up to the treaty obligations—was not the target. I could have written my letter to the

G8 leaders, but there was a better chance of gaining attention in Japan with a letter to the prime minister. Ultimately it is the public that must become informed and place pressure on our governments, which do not appreciate the lesson that Kyoto has delivered.

Before discussing the letter to Fukuda, we should consider lessons from Kyoto. The Kyoto accord, I could show, is fundamentally flawed. Yet a repeat of the Kyoto approach, with tightened caps, is exactly what the international community has been focusing on for the next international agreement. Altogether the Kyoto experience points to what not to do, but it also can help us define a more basic, workable approach.

The Kyoto accord was doomed before it started, because it did not attack the basic problem. The Kyoto Protocol set emissions reduction targets for developed countries, with targets negotiated individually with each country. Developing countries were not required to reduce emissions, but the Kyoto accord attempted to reduce the growth of their emissions through the Clean Development Mechanism, which allowed industrialized countries to make investments aimed at reducing the growth rate of developing countries' emissions in lieu of their own.

One flaw in the Kyoto approach is that targets for emissions reduction do not work, regardless of whether they are voluntary or "legally binding." Most countries demand concessions or a favorable target before they will agree to a target, or before they will ratify an accord. Russia, for example, agreed to the Kyoto accord only after it became clear they could achieve their target without serious effort, and thus they would be able to sell emission allowances

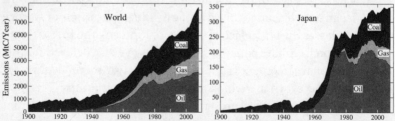

FIGURE 25. *Fossil fuel emissions by fuel type for the world and Japan. (Data sources are Carbon Dioxide Information Analysis Center, Oak Ridge National Laboratory, and British Petroleum.)*

to countries that failed to meet their own target. Another problem with targets is that there is no good way to prevent a country from overshooting its target, as many countries did.

Offsets are another of Kyoto's flaws. If a country finds that it is too inconvenient to meet its carbon emission reduction target, it can purchase the right to exceed its emissions target. In other words, it "offsets" its excess emissions via an action that supposedly reduces greenhouse gas emissions someplace on the planet. *Supposedly.* But only rarely do offsets actually cancel the climate effect of an overshoot in emissions, and the offset approach can even have a reverse effect. The offsets often occur, or supposedly occur, in developing countries, which will sell offsets at a low price. The existence of offsets discourages developing countries from improving their energy efficiency or reducing their greenhouse gas emissions, because the more emissions they have, the more actions they will have to sell as offsets to developed countries under the Clean Development Mechanism. Entrepreneurs in China even produced chlorofluorocarbons solely so they could be sold to developed countries and destroyed, thus providing bogus or imaginary offsets. Other offsets are practically unverifiable or temporary, for example, tree plantings in developing countries. But even if the tree planting is legitimate, I showed in chapter 8 that reforestation is not an alternative to fossil fuel phaseout—*both* actions are required to get the carbon dioxide amount back below 350 ppm.

Now let's look at the empirical evidence for how well the Kyoto Protocol worked. And let's look at Japan specifically, as well as the global result. Japan's success should be a poster child for the rest of the world. It is well known that Japan works very hard at energy efficiency; it may be the most energy efficient nation in the world, at least in industrial processes. Japan has good reason to minimize its fossil fuel use, because it has little indigenous fossil fuel supply.

Figure 25 shows fossil fuel carbon dioxide emissions, by fossil fuel type, for the world and Japan. Japan did not meet its Kyoto emissions reduction target, even with the help of offsets—instead, its fossil fuel emissions increased. Global emissions skyrocketed. There was a substantial reversion to coal—the oldest, dirtiest fossil fuel.

That was the world's experience with the Kyoto Protocol. It was not a success, despite the fact that many countries suffered considerable pain in trying to meet its obligations.

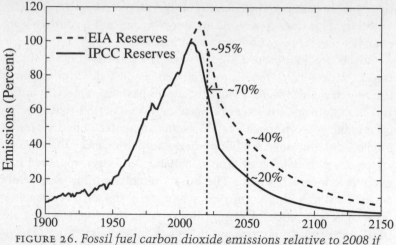

FIGURE 26. *Fossil fuel carbon dioxide emissions relative to 2008 if coal emissions are phased out over the 2010–2030 period and unconventional fossil fuels are not developed. The larger EIA oil and gas reserve estimate reflects aggressive exploitation of potential reserves. (Data from Hansen et al., "Target Atmospheric CO₂." See sources for chapter 8.)*

Today we are faced with the need to achieve rapid reductions in global fossil fuel emissions and to nearly phase out fossil fuel emissions by the middle of the century. Most governments are saying that they recognize these imperatives. And they say that they will meet these objectives with a Kyoto-like approach. Ladies and gentlemen, your governments are lying through their teeth. You may wish to use softer language, but the truth is that they know that their planned approach will not come anywhere near achieving the intended global objectives. Moreover, they are now taking actions that, if we do not stop them, will lock in guaranteed failure to achieve the targets that they have nominally accepted.

How can we say that about our governments? How can we be so sure? We just have to open our eyes. First, they are allowing construction of new coal-fired power plants. Second, they are allowing construction of coal-to-liquids plants that will produce oil from coal. Third, they are allowing development of unconventional fossil fuels such as tar sands. Fourth, they are leasing public lands and remote areas for oil and gas exploration to search for the last drop of hydrocarbons. Fifth, they are allowing companies to lease land for hy-

draulic fracturing, an environmentally destructive mining technique
to extract every last bit of gas by injecting large amounts of water
deep underground to shatter rocks and release trapped gas. Sixth,
they are allowing highly destructive mountaintop-removal and long-
wall coal mining, both of which cause extensive environmental
damage for the sake of getting as much coal as possible. In long-wall
mining, a giant machine chews out a coal seam underground—
subsequent effects include groundwater pollution and subsidence of
the terrain, which can damage surface structures. And on and on.

Can we quantify the duplicity of our governments? Can we show
that the goals for future emissions reductions are figments of their
imagination, entirely inconsistent with the policies that they are
busy adopting? Indeed we can. Figure 26 shows global fossil fuel carbon dioxide emission relative to emissions in 2008, *under the assumption* that coal emissions will be phased out linearly during
2010–2030 and that unconventional fossil fuels will not be developed. It is assumed that oil and gas will follow the usual bell-shaped
depletion curves, with two different estimates for the size of remaining reserves. The larger (EIA) reserve estimate (see figure 22) corresponds to the case in which we aggressively pursue every last drop.

Figure 26 shows that if coal emissions are phased out entirely
and unconventional fossil fuels are prohibited, fossil fuel emissions
in 2050 will be somewhere between 20 and 40 percent of emissions
in 2008. In other words, the reserves of conventional oil and gas are
already enough to take emissions up to the maximum levels that
governments have agreed on. The IPCC estimate, in which we ex-
ploit only the most readily available oil and gas, allows the possibility of getting emissions levels back to 350 ppm this century.

The problem is that our governments, under the heavy thumb of
special interests, are not pursuing policies that would restrict our
fossil fuel use to conventional oil and gas and move the world rapidly toward a post-fossil-fuel economy. Quite the contrary, they are
pursuing policies to get every last drop of fossil fuel, including coal,
by whatever means necessary, regardless of environmental damage.
With the policies governments are pursuing, fossil fuel emissions
will be much larger in 2050 than shown in figure 26, and possibly
larger than emissions today.

I emphasize, first, that a linear phaseout of coal emissions by 2030
(emissions reduced to half by 2020) is a huge challenge, requiring

urgent actions now. Developed countries will need to complete their coal phaseout by about 2020. That is a tall order. For example, the United States obtains half its electricity from coal-fired power plants, all of which will need to be replaced by some combination of improved energy efficiency and alternative energy sources. The fact that developed countries are not scheduling a rapid phaseout of coal plants and working hard on alternatives shows that, in fact, they have no realistic expectation of meeting their stated goals.

I emphasize, second, that we have all the ingredients we need to meet this challenge—except leadership willing to buck the special financial interests benefiting from business as usual. The tragic aspect of this story is that the specific actions that we have so far neglected to take—which I will describe momentarily—would actually have great benefits for the nation, for nature, for our children and grandchildren.

"Foo," you may be thinking. "This must be exaggerated nonsense. Our leaders are not so stupid that they would turn their backs on sensible policies with multiple benefits." Hold on. Here are a few brief comments to try to allay your concerns about implausibility. Think Washington. Think lobbyists. Think revolving doors. There were 2,340 registered energy lobbyists when I checked in early 2009 (more now, and not all are registered, by any means). As an example, one lobbyist, former House Democratic leader Dick Gephardt, received $120,000 from coal company Peabody Energy in 2008—per quarter. That's almost half a million dollars per year. As they say back in my hometown, "It's good pay if you can get it."

My talk at the United Nations University, coincidentally, was also on our Independence Day, July 4, 2008. I finished my letter to Prime Minister Fukuda on the bus from Narita airport to Tokyo on July 3. The university kindly arranged for interviews with the media that afternoon and for printing of the letter. The next morning the university hand-delivered the letter to the prime minister's office. At first the prime minister's assistant would not accept the letter—until he was informed that it was the subject of an article in the *Mainichi Daily News*, one of the leading Tokyo newspapers.

The letter to Prime Minister Fukuda (available on my Web site) was more policy-specific than most of the letters I had written to other national leaders and U.S. governors over the preceding year. The letter was intended for a broad audience—the G8 leaders and the public—rather than only the prime minister. I may have been feeling a bit frustrated, as the final two paragraphs, added during the bus ride from Narita, were probably a bit cheeky and perhaps even sarcastic:

> Finally, Prime Minister Fukuda, I would like to thank you for helping make clear to the other leaders of the eight nations the great urgency of the actions needed to address climate change. Might I make one suggestion for an approach you could use in drawing their attention? If the leaders find that the concept of phasing out all emissions from coal, and taking measures to ensure that unconventional fossil fuels are left in the ground or used only with zero-carbon emissions, is too inconvenient, then, in that case, they could instead spend a small amount of time composing a letter to be left for future generations.
>
> The letter should explain that the leaders realized their failure to take these actions would cause our descendants to inherit a planet with a warming ocean, disintegrating ice sheets, rising sea level, increasing climate extremes, and vanishing species, but it would have been too much trouble to make changes to our energy systems and to oppose the business interests who insisted on burning every last bit of fossil fuels. By composing this letter the leaders will at least achieve an accurate view of their place in history.

My experiences in the U.K., Germany, and Japan are representative. My correspondence with other governments, notably Australia, and with several U.S. governors is available on my Web site. Most of the politicians advertised themselves as being "green," but what I learned was that, invariably, it amounted to greenwash, demonstrating token environmental support while kowtowing to fossil fuel special interests. To be generous, most of these leaders probably kidded themselves into believing that their modest green efforts were meaningful.

There are rays of hope, however. There seems to be a chance that the U.K. will phase out coal emissions by 2025. However, this progress was not a result of persuasion due to the scientific rationale that I presented in visits to the U.K. Instead it was based on a popular campaign there in which citizens—especially activists, but also the mainstream public, scientists, the financial sector, and even some politicians—exposed government greenwash. I will discuss such activities further in chapter 11.

Real World Data: Evaluating What Works

Okay, given that our political leaders do not want to face up to the truth, what do we do? The worst thing would be to stick our heads in the sand and let the politicians and fossil fuel industry get away with their short-term views and their short-term profits, allowing them to destroy the prospects for young people and future generations. Instead, we need to look at the situation objectively and strategically, ask what the world must aim for, examine empirical data that can help us evaluate what works and what does not—and then even throw in some common sense.

Let's return to some empirical data from way back in figure 2 (chapter 2, page 21). The graph shows that most of the energy in the United States comes from fossil fuels. It also shows Amory Lovins's target for replacing all fossil fuels with renewable energy—his scenario eliminates nuclear power and large hydroelectric power as well. The good news shown in figure 2 is that improved energy efficiency during the past three decades reduced the expected growth of U.S. energy use, even though the population increased. The bad news is that the rapid ascendancy that Lovins foresaw for "soft" renewable energies, such as the wind and sun, never occurred—their contribution to total U.S. energy is still minuscule.

Why have fossil fuels continued to reign supreme? The main reason is price—fossil fuels are cheap. Oil and gas are also convenient, portable fuels. And fossil fuels are a reliable source of electricity, not intermittent like the wind and sun. Governments have tried to spur renewable energies by requiring utilities to obtain a specified fraction of their electricity generation from renewable sources, but success so far has been limited. Moreover, some people argue that the renewable energies are not so "soft"—arrays of solar collectors and wind turbines have their own environmental footprint, and often require new power lines to carry electricity from remote areas to population centers.

Germany provides useful empirical evidence about progress in quitting the fossil fuel addiction. Germany is making a major effort

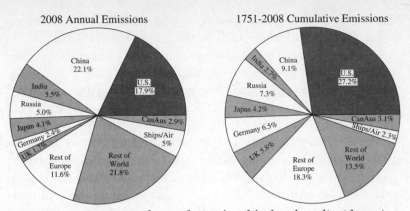

FIGURE 27. *Current and cumulative fossil fuel carbon dioxide emissions. (Data are updates of Hansen et al., "Dangerous Human-made Interference with Climate." See sources for chapter 7.)*

to improve energy efficiency. It is also trying hard to promote renewable energy, with large subsidies for wind and solar energies. Wind provides up to 20 percent of the country's electric energy in winter, but on annual mean the wind and sun produced only 7.3 percent of Germany's electricity in 2008. That renewable fraction is still growing, but at a cost—some industries have cited increased electric rates as a reason for relocating outside Germany.

But what is disturbing about the empirical evidence from Germany is that, despite technical prowess and strong efforts in energy efficiency and renewable energies, there are no plans to phase out coal. On the contrary, there are plans to build new coal-fired power plants, which the German government claims will be necessary once the country closes its nuclear power plants. The bottom line seems to be that it is not feasible in the foreseeable future to phase out coal unless nuclear power is included in the energy mix.

Other important empirical facts concern where carbon dioxide emissions are coming from now and who is responsible for the burden of fossil fuel carbon dioxide that has accumulated in the atmosphere. The left chart in figure 27 shows that China has passed the United States as the country with the largest current rate of carbon dioxide emissions and that India is now third, behind the United States. However, *cumulative* carbon dioxide emissions are the proper measure of responsibility for human-caused climate change, and, as shown by the chart on the right in figure 27, the United

States has a responsibility about three times that of China. European responsibility is about the same as that of the United States.

One more critical set of empirical facts: Coal accounts for three quarters of the carbon dioxide emissions of both China and India, and their coal use is mainly for electricity generation. Coal accounts for only 20 to 40 percent of carbon dioxide emissions among all other major emitting nations, with the United States being at the high end of that range (i.e., about 38 percent). China and India are not only the first and third greatest emitters of carbon dioxide; their emissions are also growing the fastest among the major emitting countries. Any solution to the global warming problem must address the electrical energy needs of China and India.

Now, what are the means by which fossil fuel use can be reduced and eventually phased out? The first priority, everybody agrees, must go to energy efficiency. There is great potential for energy savings at little cost or even with financial benefit. People in the United States, Canada, and Australia use about twice as much energy per capita as those in Europe or Japan, where the quality of life and gross domestic product per capita are just as high. Contrary to widespread belief, only a small part of the difference in energy use is accounted for by greater travel distances in the United States, Canada, and Australia. The primary difference is because Europe and Japan have taken steps to minimize fuel needs. Higher taxes on fossil fuels, equivalent to several dollars per gallon of gasoline, provide a strong disincentive against inefficient vehicles in Europe and Japan.

Are similar efficiencies possible in the United States? California achieves energy efficiency close to that of Europe and Japan. Since 1975, per capita use of electricity in California has remained con-

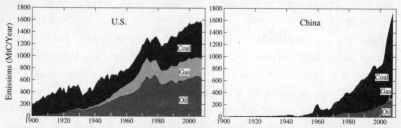

FIGURE 28. *Fossil fuel emissions by fuel type for the United States and China. (Data sources are Carbon Dioxide Information Analysis Center, Oak Ridge National Laboratory, and British Petroleum.)*

stant, while growing 50 percent in the rest of the United States. The reason is that California has an astounding variety of energy efficiency standards and incentives, along with higher electric rates that "internalize" the idea of conservation at the individual level. Utility regulations in California also are structured such that the utilities make more money by encouraging efficiency rather than by selling more energy.

The second priority, behind energy efficiency, is renewable energies, and again this has widespread agreement. Governments can encourage renewables via tax incentives and "renewable portfolio standards," a requirement that utilities use renewable energy for some fraction of their power generation. Still, ample experience demonstrates that governments should not try to pick the technology "winners," choosing to support specific technologies with taxpayer funds; they will waste a lot of our money if they do. Instead, the most effective policy would be to remove subsidies for fossil fuels and add a flat carbon tax on emissions, as I will discuss.

Will energy efficiency and renewables be enough? That is the assertion of Amory Lovins and perhaps a few others, but most energy experts believe that Lovins is overly, even wildly, optimistic. Of course we should do everything practical to help efficiency and renewables cover as much of our energy needs as possible. And we can hope that at some point in the future efficiency and renewables will be able to satisfy all energy needs in the United States. But that is not going to happen in the next decade or two, a conclusion that is true a fortiori in China and India.

How can we be so certain that efficiency and renewables cannot quickly eliminate the need for fossil fuels? Didn't Al Gore propose, in 2008, that the United States could have carbon-free electric energy in ten years? That transformation would require trillions of dollars and a government project comparable in scale to Franklin Roosevelt's Works Progress Administration, but with skilled labor. There is no sign of this happening or even being proposed by the Obama administration. We have another way to gauge this matter, using empirical data: We can look at a high-tech nation, Germany, which has one quarter of the population of the United States and an energy requirement about ten times less than the United States, and which has gone to great effort and cost to spur renewables energy and efficiency over the past decade. The result, in a nation with exceptional technical prowess and determination: Germany has barely made a dent in its carbon dioxide emissions and is planning more coal plants.

Now add China and India to the equation. You can see in figure 28 how rapidly China's emissions have increased. The growth rates of India's energy use and carbon dioxide emissions are comparable to those of China. The population of India, approaching 1.3 billion, is expected to exceed that of China by about 2025. The energy needs of China and India will continue to grow as both nations work to raise the living standards of their populations.

Energy efficiency and renewable energies should be top priorities in China and India, just as in the West. But my point is this: Efficiency and renewables are not going to be sufficient for their energy needs during the next several decades. That is probably true in the rest of the world also—as the examples of Germany and Japan, countries that are trying hard, illustrate—but China and India make the conclusion undeniable. We must rapidly increase sources of carbon-free energy if we are to solve the climate problem.

At this point some people throw up their hands in despair, concluding that because there is no hope that China and India will curtail their carbon dioxide emissions, the planet is therefore doomed. That knee-jerk assessment, I am confident, is wrong. Why?

First, China and India would suffer enormously if the climate is allowed to spiral out of its Holocene range. India does not want 100 million Bangladeshi refugees on its doorstep. India itself has more than 100 million people living near sea level. China has 300 million people living within a twenty-five-meter elevation of sea level. China's long history under reasonably stable climate patterns provides them a heritage that they will want to protect—and they'll also want to avoid severe disruption from rising sea level, shifting climate patterns, loss of mountain glaciers, and intensifying floods and droughts.

Second, the Chinese and Indian cultures respect science, and their governments are capable of moving promptly in response to national needs. China is already making major investments in energy efficiency and renewable energies. I have organized two climate workshops, both concerning air pollution and climate change, at the East-West Center in Hawaii that included a number of Chinese and Indian participants. All indications are that the scientists and national leaders appreciate what the science is revealing, are positive about international cooperation, and are eager to find ways to clean up air and water pollution while continuing economic development.

However, what China and India require—indeed, what is needed in most countries—to phase out coal emissions is a carbon-free source of baseload electric power that is competitive in price with coal. "Baseload" means sources of electric power capable of continuous operation, unlike current capabilities of wind and sun power.

I do not mean to denigrate the potential for renewable energies to provide continuous power. There are always some places where the sun is shining or the wind is blowing, there are developing concepts for large-scale "batteries" to store wind and solar energy, and expanded low-loss electric grids can connect widespread areas to move energy from where it is available to where it is needed. Such energy storage and long-distance energy transfer have cost and environmental impacts, but renewable energies should be used to the degree that is practical.

However, most energy experts agree that, for the foreseeable future, renewable energies will not be a sufficient source of electric power. There is also widespread agreement that there are now just two options for nearly carbon-free large-scale baseload electric power: coal with carbon capture and storage, and nuclear power. Let's consider the problems with each of these options.

Clean Coal?

Capture of carbon dioxide at power plants appears to be technically feasible. The carbon dioxide then needs to be piped to a location where it can be injected far underground, to a sufficient depth in a geologic setting where it is not expected to escape. The capture process takes energy—an enormous amount—so about 25 percent more coal must be burned to add the capture option. If it must be piped a significant distance, that adds more cost. There is likely to be a NIMBY (not in my backyard) reaction of the public to proposed sites for carbon dioxide burial, since a large-scale escape of the gas would be dangerous—carbon dioxide can suffocate humans, as it did residents near an African lake from which a natural pocket of carbon dioxide escaped in 1986.

Carbon capture and storage may be a viable approach in some countries. There are several nations now developing power plants with carbon capture and storage. But it is implausible to think a developing nation such as India would replace its existing huge number of coal-fired power plants—unless the West were willing to pay the cost differential. Such a demand would be reasonable, because the per capita Indian contribution to the climate problem is about a factor of twenty less than that of the industrial West, as shown in figure 24.

The bill for the West? Trillions of dollars for new carbon-capturing power plants to replace all the old ones in China and India that emit carbon dioxide to the atmosphere. Then there is the increased operating costs of plants that capture and store carbon dioxide. Who will pay the added costs? The West? You get the idea. This is not going to happen. Coal plants with carbon capture and storage are not going to happen on a large scale in the West either. There are countries saying that they will build power plants that are "carbon capture ready." They are misleading you. The politicians know that the public, at least in most countries, will never accept the large increases in electricity price that would accompany carbon capture, let alone accept burial of the carbon dioxide in their neighborhood.

Besides, what about the mercury, arsenic, sulfates, and other air and water pollutants that come with coal? They can be reduced with capture, but not eliminated. And the problems at the mines, especially the horrendous mountaintop-removal and long-wall mining? Can we not move on from this cursed remnant of the first phase of the industrial revolution?

Coal companies are spending huge amounts to put lipstick on coal, but it is hard to hide the fact that it is pretty ugly stuff. Well, then, what about the other extant option for large-scale carbon-free baseload electricity—nuclear power?

Atoms for Peace?

No new nuclear power plants have been ordered and put into operation in the United States in more than thirty-five years, since well before the Three Mile Island accident at a nuclear power plant in Pennsylvania on March 29, 1979. A combination of design flaws and inadequate control room procedures caused a partial meltdown of the reactor core and the release of a small amount of radioactive gases to the atmosphere. President Jimmy Carter's accident investigation board, the Kemeny Commission, headed by Dartmouth College president John Kemeny, cast blame for the accident widely: The reactor manufacturer and the utility company operating the plant were criticized for lapses in quality assurance, maintenance, and operator training and for failing to define clear operating room procedures. And the Nuclear Regulatory Commission was blamed for inadequate oversight.

Chances of having another accident like that at Three Mile Island have been reduced via better operating procedures and oversight, but the public is justifiably skeptical about the ability to eliminate human lapses. A more fail-safe reactor design seems essential to achieve broad public acceptance. Fortunately, human impact from the accident was small. The Kemeny Commission determined that "there will either be no case of cancer or the number of cases will be so small that it will never be possible to detect them. The same conclusion applies to other possible health effects." Several subsequent studies have been unable to find any significant health effects from the accident, but issues about the adequacy of the data are still debated. A useful illustration of the health risk posed by the Three Mile Island accident is that the increased exposure to radioactivity suffered by people living nearby is comparable to the radiation exposure that people receive in a chest x-ray or in a round-trip transcontinental flight at the altitude at which commercial jets fly.

Nevertheless, a strong enduring negative public reaction against nuclear power ensued, probably in part because of a fortuitous event: Twelve days before the Three Mile Island accident, a popular movie called *The China Syndrome* opened. The movie, starring Jane Fonda, concerned an accident at a nuclear power plant. The real accident in Pennsylvania enhanced the credibility of antinuclear activists and engendered mass antinuclear demonstrations. One rally in New York, with speeches by Jane Fonda and Ralph Nader, drew two hundred thousand people and was followed by a series of nightly "No Nukes" concerts in Madison Square Garden. Another rally, in Washington, drew sixty-five thousand people, including the governor of California.

Then a much more serious nuclear accident occurred in Chernobyl in the Soviet Union in 1986. Unlike most Western nuclear power plants, most early Soviet reactors had no hard containment vessel. As a result, a huge cloud of radioactive material was spread over large areas of the Soviet Union and Europe after a steam and chemical explosion blew apart one of the Chernobyl reactors. The World Health Organization calculates that there might be as many as four thousand cancer deaths because of radiation released at Chernobyl, which compares with one hundred thousand other cancer deaths among the same population.

There are several serious issues with nuclear power, which I will soon note. But to be objective, the empirical data on the human consequences of the early nuclear power plants should be compared with data on the consequences of coal use.

Leading world air-pollution experts at our workshops at the East-West Center in Honolulu agreed that there are at least one million deaths per year from air pollution globally. It is difficult to apportion the deaths among different pollution sources—such as vehicles and power plants—because people are affected simultaneously by all sources. But to get an idea of the numbers, let's first assign 1 percent to coal-fired power plants. That's ten thousand deaths per year—every year.

Actually, all experts agree that coal is responsible for far more than 1 percent of the air pollution. In fact, recent data show that more than 1 percent of some air pollutants in the United States comes from *Chinese* power plants! I point this out to emphasize that pollution and climate change are global problems—we must work together with other countries to solve them. Assigning 10 percent of global air pollution deaths to coal is probably still conservative—that's a hundred thousand deaths per year, every year.

Yet there are no two-hundred-thousand-person rallies against coal, no nightly "No Coal" concerts. Death by coal is probably not as sexy as death by nuclear accident. Perhaps we have greater fear of nuclear power because it is more mysterious than that familiar black lump of coal—even though we know coal contains remarkably bad stuff.

When asked about nuclear power, I am usually noncommittal, rattling off pros and cons. However, there is an aspect of the nuclear story that deserves much greater public attention. I first learned about it in 2008, when I read an early copy of *Prescription for the Planet*, by Tom Blees, who had stumbled onto a secret story with enormous ramifications—a story that he delved into by continually badgering some of the top nuclear scientists in the world until he was able to tell it with a clarity that escapes technical experts. I have since dug into the topic a bit more and observed how politicians and others reacted to Blees's information, and the story has begun to make me slightly angry—which is difficult to do, as my basic nature is very placid, even comfortably stolid.

Today's nuclear power plants are "thermal" reactors, so-called because the neutrons released in the fission of uranium fuel are slowed down by a moderating material. The moderating material used in today's commercial reactors is either normal water ("light water") or "heavy water," which contains a high proportion of deuterium, the isotope of water in which the hydrogen contains an extra neutron. Slow neutrons are better able to split more of the uranium atoms, that is, to keep nuclear reactions going, "burning" more of the uranium fuel.

The nuclear fission releases energy that is used to drive a turbine, creating electricity. It's a nice, simple way to get energy out of uranium. However, there are problems with today's thermal nuclear reactors (most of which are light-water reactors). The main problem is the nuclear waste, which contains both fission fragments and transuranic actinides. The fission fragments, which are chemical elements in the middle of the periodic table, have a half-life of typically thirty years. Transuranic actinides, elements from plutonium to nobelium that are created by absorption of neutrons, pose the main difficulty. These transuranic elements are radioactive material with a lifetime of about ten thousand years. So we have to babysit the stuff for ten thousand years—what a nuisance that is!

Along with our having to babysit the nuclear waste, another big problem with thermal reactors is that both light-water and heavy-water reactors extract less than 1 percent of the energy in the original uranium. Most of the energy is left in the nuclear waste produced by thermal reactors. (In the case of light-water reactors, most of the energy is left in "depleted-uranium tailings" produced during uranium "enrichment"; heavy-water reactors can burn natural uranium, without enrichment and thus without a pile of depleted-uranium tailings, but they still use less than 1 percent of the uranium's energy.) So nuclear waste is a tremendous waste in more ways than one.

These nuclear waste problems are the biggest drawback of nuclear power. Unnecessarily so. Nuclear experts at the premier research laboratories have long realized that there is a solution to the waste problems, and the solution can be designed with some very attractive features.

I am referring to "fast" nuclear reactors. Fast reactors allow the neutrons to move at higher speed. The result in a fast nuclear reactor is that the reactions "burn" not only the uranium fuel but also all of the transuranic actinides—which form the long-lived waste that causes us so much heartburn. Fast reactors can burn about 99 percent of the uranium that is mined, compared with the less than 1 percent extracted by light-water reactors. So fast reactors increase the efficiency of fuel use by a factor of one hundred or more.

Fast reactors also produce nuclear waste, but in volumes much less than slow (thermal) reactors. More important, the radioactivity becomes inconsequential in a few hundred years, rather than ten thousand years. The waste from a fast reactor can be vitrified—transformed into a glasslike substance—placed in a lead-lined steel casket, and stored on-site or transported for storage elsewhere. Plus, this waste material cannot be used to make explosive weapons (although it could be used in a "dirty bomb," which is best described as a weapon of mass disruption, rather than mass destruction, because it can do relatively little physical damage).

"Wait a minute!" you may be thinking. "If there is a type of nuclear power that is so good, how come nobody knows about it?" Let me tell that story.

The concept for fast-reactor technology was defined by Enrico Fermi, one of the greatest physicists of the twentieth century and a principal in the Manhattan Project, and his colleagues at the University of Chicago in the 1940s. By the mid-1960s the nuclear scientists at Argonne National Laboratory had demonstrated the feasibility of the concept. The nuclear experts, through the Department of Energy chain of command, informed political leaders about the situation. The leaders got the message.

Richard Nixon, in his June 4, 1970, presidential energy message to

Congress, said, "Our best hope today for meeting the nation's growing demand for economical clean energy lies with the fast breeder
reactor." The highest priority of the energy program, he announced,
should be a "commitment to complete the successful demonstration
of the liquid-metal fast breeder reactor." The Joint Committee on
Atomic Energy of Congress concurred with this goal.

By the way, Nixon used the adjective "breeder" because fast reactors can be run such that they produce more nuclear fuel than they
consume. They are not creating energy out of nothing; they are just
converting "fertile" elements into a fuel that is directly usable in a
reactor, i.e., into "fissile" elements—elements that are fissionable
when hit by a slow (thermal) neutron. It is necessary to supply a fast
reactor with "fertile" material, but there is enough of that available
in the nuclear waste piles that we are babysitting to last many centuries. Fertile material that can be burned in fast reactors is contained in by-products of past weapons development programs as
well as in the waste piles from light-water reactors. The United
States is presently storing about six hundred thousand tons of uranium hexafluoride, a by-product of nuclear weapons production. A
reasonable assessment of the value of this material as fuel, if fast reactors were deployed as the energy source for power plants, is about
$50 trillion. Yes, trillion. But it will take almost a thousand years to
use all that fuel, so don't expect a customer to buy it all at once.

"Liquid metal" refers to the coolant used in the reactor. The
usual choice for the metal is sodium, which is liquid over a wide
range of temperature (between 98 and 883 degrees Celsius). Liquid
metals have a safety advantage over water, because they do not
need to be kept under pressure, and liquid sodium is noncorrosive.

Nixon thought that fast reactors would be providing most of our
electricity in the twenty-first century. What happened? Three Mile
Island, for one thing. All nuclear power was lumped into one bag, a
fearsome one. Substantial "antinuke" sentiment developed. Several
environmental groups came out strongly against nuclear power.
Most of the public was not adamantly opposed to it, but nuclear
power's contribution to U.S. electricity stopped growing, stabilizing at about 20 percent, with fossil fuels providing most of the remainder.

The Department of Energy kept nuclear power research alive.
The United States had the top nuclear experts in the world, and the

top laboratory was Argonne National Laboratory. A low level of support allowed steady progress to continue until, in 1994, the Argonne scientists had tested all the necessary components and were ready to build a demonstration fast-reactor power plant. At that point, the Clinton-Gore administration canceled the program. In his 1994 State of the Union address, Bill Clinton announced, "We will terminate unnecessary programs in advanced reactor development."

That was not a rational decision in my opinion. It is hard to understand it on a scientific basis. To my mind, the most likely interpretation is that the antinuke people got worried that this next-generation nuclear power was getting too close to becoming a reality. Strange as it may seem, I doubt that Clinton and Gore, who were well aware of global warming, did an in-depth analysis of this potential energy source. At meetings of heads of state, Clinton was often described, probably accurately, as "the smartest guy in the room," but he never seemed to take a great interest in "details" about energy. As Tom Blees points out in his book, Clinton had used antinuclear sentiment in the Democratic Party to his advantage in the 1992 primaries, describing an opponent as "pro-nuclear," as if that were patently stupid. So perhaps it is not surprising that Clinton's secretary of energy, Hazel O'Leary, terminated the research, either on her own or at Clinton's direction. It was a clean kill: Argonne scientists were told not only to stop the research but also to dismantle the project—and those who had worked on the project were instructed by the DOE to *not* publicize it. In congressional debates Senator John Kerry was the principal bearer of the antinuclear flag. That may explain why Gore, when questioned about the 1994 decision on the floor of the United States Senate in 2008, had a quizzical look, as if he could not remember. It seems possible that antinuke people, who heavily support the Democratic Party, were being repaid, without a whole lot of analysis.

That 1994 decision, whether driven by politics or not, is water under the bridge. What is the sensible thing to do now? At the very least, we should build a test fast-reactor nuclear power plant. The fast-reactor approach is sometimes called fourth-generation nuclear power. The existing light-water nuclear power plants in the United States (there are about one hundred) are the second generation. Third-generation nuclear power plants are the ones that industry is proposing now, with several in the approval process. These third-generation nuclear plants are thermal reactors, mostly light-water reactors, but with an improved design to simplify operations and increase safety.

If there is to be a nuclear renaissance in the United States, it will be led by third-generation nuclear power plants, which are ready to go now. However, a substantial nuclear renaissance, able to supplant a large portion of coal power, will occur only if we are confident that fourth-generation power plants are on the way. The fourth-generation plants are needed to deal with the nuclear waste from the third-generation plants and to meet growing energy demand in the future.

Apprehension about nuclear power will diminish as word spreads that the nuclear energy safety record, already unsurpassed by any comparable industry, is even further enhanced by the advent of reactors that will shut down harmlessly, without human intervention, in response to abnormal situations such as those that triggered the accidents at Three Mile Island and Chernobyl.

The combination of third- and fourth-generation nuclear power plants would solve another problem sometimes cited by foes of nuclear power: uranium mining. They claim we will run out of uranium in several decades and also that it takes a lot of energy to mine and process uranium. With fourth-generation nuclear power plants in the mix, that debate disappears. We already have enough fuel stockpiled, in nuclear waste and by-products of nuclear weapons production, to supply all of our fuel needs for about a thousand years.

In fact, given that fast reactors make it economical to extract uranium from seawater, we now have enough fuel, in theory, to run nuclear power plants for several billion years. In other words, nuclear fuel is inexhaustible, putting it in the same category as renewable solar energy.

Another concern about nuclear power, in addition to nuclear waste, is the possibility of weapons-grade nuclear material falling into the hands of terrorists or rogue nations. Weapons proliferation is a valid concern, and a serious one, but the danger is not increased by fourth-generation nuclear power. Many nuclear opponents seem to believe that this weapons danger will be removed or at least reduced if nuclear power is eliminated in the United States. On the contrary. The nuclear genie is out of the bottle. And several nations are already working on fourth-generation reactors.

Rather than the United States abandoning advanced reactor development, the better approach would be for the United States to (once again) lead that technology development—in a direction that minimizes proliferation risks. Furthermore, it is difficult to see how the international organizations to control proliferation can be effective without U.S. leadership—and how can we lead, if we abandon the technology? The way to minimize nuclear proliferation risks is to be a leader in the technology, to make it as fail-safe and proliferation-resistant as possible, and to cooperate in international management of nuclear material.

When I became acquainted with this matter in 2008, I began recommending in public talks that the United States should initiate urgent development of a demonstration fourth-generation nuclear power plant. There would be no need to decide immediately about commercialization of fourth-generation technology, but we should understand its potential. Indeed, that knowledge affects the viability of third-generation nuclear power plants—can we anticipate help from fourth-generation technology to solve the nuclear waste problem?

Urgency derives from the need for a feasible way to phase out coal rapidly. If energy efficiency and renewable energy can handle all future energy needs, that would be great. But it is extremely irresponsible, in my opinion, to make the *assumption* that efficiency and renewables are all that will be needed.

I have spoken with numerous utility executives and their technical staff. Every one of them asserts that efficiency and renewables will not be enough in the foreseeable future. In practice, they say, they will need either fossil fuels or nuclear power for baseload capacity. Maybe they are wrong—maybe they are underestimating the potential of efficiency and renewables—but it would be foolish for us to assume that they are all wrong. Rather, it seems clear that efficiency and renewable energies will not be sufficient to allow phaseout of coal.

Yet, when I recommended urgent testing of fourth-generation nuclear power capabilities, I was bombarded with messages from environmentalists and antinuclear people. Mostly it was friendly advice—after all, they agreed with my climate concerns—but they invariably directed me to one or more of a handful of nuclear experts. Some of the experts were associated with organizations such as the Natural Resources Defense Council, the World Wildlife Fund, or the Union of Concerned Scientists—and there was Amory Lovins of Rocky Mountain Institute.

These are fine organizations. I am sure that I agree with more than 90 percent of the things that they advocate, and I am proud to have received the World Wildlife Fund's Duke of Edinburgh Conservation Medal from Prince Philip himself. (Anniek and I got to have lunch with him, but he would not answer a question that I brought from my granddaughter: What is your favorite color? He said that if he specified a favorite color, all future gifts would be that color.)

Then I learned that the same small number of organizations and experts, who had been repeating the same message for decades, had an influence way out of proportion to their numbers. I found that members of Congress and their staffers, none of them nuclear scientists, were getting most of their advice on nuclear power from the same organizations. The organizations trot out the same few "experts," who speak with technical detail that snows the listener and who conclude that the United States, in effect, should terminate peaceful use of nuclear energy.

That's what began to make me a bit angry. Do these people have the right to, in effect, make a decision that may determine the fate of my grandchildren? The antinuke advocates are so certain of their righteousness that they would eliminate the availability of an alternative to fossil fuels, should efficiency and renewables prove inadequate to provide all electricity. What if the utility executives are right, and we must choose between coal or nuclear for baseload power? Even if renewables are sufficient to produce the electricity needed by the United States, what about India and China? It's one world, and we have to live with pollution from China and India.

But, you may say, aren't these good scientific organizations representing a lot of people who are making recommendations based on the scientific method? Umm, not exactly. The Union of Concerned Scientists seems to me to be a lobbying organization. It lobbied hard for the Kyoto Protocol. When I published the "alternative scenario" paper, which the organization considered to be critical of the Kyoto Protocol, it encouraged the writing of an article criticizing our paper and sent out an "information update" to its members strongly criticizing and mischaracterizing our paper. The union would not provide me with its mailing list so that I could respond. So I wrote an open letter, published on the naturalSCIENCE Web site, to correct misimpressions. Since then I have referred to it, kiddingly of course, as the Union of Concerned Lobbyists. I agree with most of what it promotes, but people should understand that this is not a group of scientists in lab coats sending out their most recent scientific analysis. The head of the organization is not a scientist and neither are many of the members.

I will discuss nonprofit and environmental organizations a bit more later, because they have a huge influence on a topic that is broader and more important than nuclear power. But first, please allow me one (long, sorry!) paragraph to summarize the nuclear story.

The scientific method requires that we keep an open mind and change our conclusions when new evidence indicates that we should. The new evidence affecting the nuclear debate is climate change, specifically the urgency of moving beyond fossil fuels to carbon-free energy sources. We need an urgent, substantial research and development program on fourth-generation nuclear power, so that we have at least one viable option in the likely event that efficiency and renewables cannot provide all needed energy. A phaseout of coal emissions in the West can proceed promptly on the basis of efficiency, renewables, third-generation nuclear power, and possibly a contribution from carbon capture and storage—although it also requires a price on carbon emissions, as discussed below. A phaseout of coal emissions in China and India almost certainly requires a cost-competitive alternative to coal. One reason for urgent development of fourth-generation nuclear power is the possibility of producing a design for a modular reactor, which would reduce costs if built in large number. It is conceivable that next-generation nuclear power might begin to be broadly deployed in China or India as early as the 2020s. Deployment would be soonest if the United States would cooperate with these nations and treat this as a matter of urgency. If you do not believe that such rapid development is feasible, you should read some of the stories about the Manhattan Project.

The Main Story

We have finally arrived at the main story: what we need to do to solve the climate problem, and how we can save the future for our grandchildren.

The problem demands a solution with a clear framework and a strong backbone. Yes, I know that halting and reversing the growth of carbon dioxide in the air requires an "all hands on deck" approach—there is no "silver bullet" solution for world energy requirements. People need to make basic changes in the way they live. Countries need to cooperate. Matters as seemingly intractable as population must be addressed. And the required changes must be economically efficient. Such a pathway exists and is achievable.

Let's define what a workable backbone and framework should look like. The essential backbone is a rising price on carbon applied at the source (the mine, wellhead, or port of entry), such that it would affect all activities that use fossil fuels, directly or indirectly. Our goal is a global phaseout of fossil fuel carbon dioxide emissions. We have shown, quantitatively, that the only practical way to achieve an acceptable carbon dioxide level is to disallow the use of coal and unconventional fossil fuels (such as tar sands and oil shale) unless the resulting carbon is captured and stored. We realize that remaining, readily available pools of oil and gas will be used during the transition to a post-fossil-fuel world. But a rising carbon price surely will make it economically senseless to go after every last drop of oil and gas—even though use of those fuels with carbon capture and storage may be technically feasible and permissible.

Global phaseout of fossil fuel carbon dioxide emissions is a stringent requirement. Proposed government policies, consisting of an improved Kyoto Protocol approach with more ambitious targets, do not have a prayer of achieving that result. Our governments are deceiving us, and perhaps conveniently deceiving themselves, when they say that it is possible to reduce emissions 80 percent by 2050 with such an approach.

A simple proof of the contrary is provided by reviewing the Kyoto results. Japan is an exemplary world citizen and was the strongest promoter of the Kyoto Protocol, so quantification of its performance is informative. Japan agreed to reduce emissions 6 percent below 1990 levels, made an honest effort, and played by the rules. What was the result? In August 2009 Japan announced that its emissions *exceeded* 1990 levels by 9 percent—missing its target by 15 percent. Japan will reduce the huge gap between target and reality by purchasing offsets of 1.6 percent via the Clean Development Mechanism and 3.8 percent via funding of tree planting. Unfortunately, these offsets are not meaningful, as I will explain. But even if we count them, Japan is nowhere near its target.

The world as a whole did much more poorly than Japan, as shown in figure 25. Results fluctuated from place to place, depending on historical accidents, not on anything that the Kyoto Protocol engendered. Germany did well because it incorporated East Germany and closed down dirty, inefficient communist-era factories. The U.K. did well because North Sea gas allowed it to close most coal mines and replace coal-fired power with gas. But overall, global emissions shot up faster than Japan's.

A successful new policy cannot include any offsets. We specified the carbon limit based on the geophysics. The physics does not compromise—it is what it is. And planting additional trees cannot be factored into the fossil fuel limitations. The plan for getting back to 350 ppm assumes major reforestation, but that is *in addition to* the fossil fuel limit, not *instead of*. Forest preservation and reforestation should be handled separately from fossil fuels in a sound approach to solve the climate problem.

The public must be firm and unwavering in demanding "no offsets," because this sort of monkey business is exactly the type of thing that politicians love and will try to keep. Offsets are like the indulgences that were sold by the church in the Middle Ages. People of means loved indulgences, because they could practice any hanky-panky or worse, then simply purchase an indulgence to avoid punishment for their sins. Bishops loved them too, because they brought in lots of moola. Anybody who argues for offsets today is either a sinner who wants to pretend he or she has done adequate penance or a bishop collecting moola.

Let us return one more time to figure 2 of chapter 2 (page 21), which provides an overview of prospective actions for phasing out carbon emissions. First, this graph illustrates a mistake made by energy professionals that continues to be made today. It shows that energy use in the United States grew far more slowly than energy experts predicted. Growth of energy use was moderate despite strong economic growth and an unexpectedly rapid population expansion fueled by immigration. For one decade, beginning in the late 1970s, energy use did not even increase, as a consequence of imposed improvements on vehicle fuel efficiency, escalating energy prices in the wake of the second "oil shock," and widespread cost overruns in the electricity sector.

Nevertheless, be prepared for energy experts telling you that a kazillion units of energy will be needed in 2050 or 2100. They will calculate how many square miles of solar power plants must be built every day or how many nuclear power plants must be built every year, and then they will wring their hands and perhaps try to sell you something. Yes, energy use is going to increase—mainly because parts of the world are developing rapidly and raising their standards of living and energy use. But energy growth need not be exceedingly rapid—figure 2 shows that energy use hardly grew during rapid economic growth in the world's largest economy, even though the great potential of energy efficiency was barely tapped. Also remember that the solution to the climate problem requires a phasedown of carbon emissions, not necessarily a phasedown of energy use. We will need to slow the energy growth rate and decarbonize our energy sources to solve the problem.

However, the growth rate of energy use is an important aspect of the problem, and we can gain further insight from the U.S. energy consumption curve (figure 2). The U.S. population has increased 50 percent since 1975, but energy use per person has not increased. The United States actually could have achieved much greater energy efficiency over this period, but there has been little economic incentive to do so since energy costs have been declining in real terms, or as a fraction of a person's budget. People are happy to drive gas-guzzlers when gasoline is cheap. Vehicle fuel-efficiency requirements were increased in 2009 in the U.S. by about 30 percent, to 35.5 miles per gallon, the only increase since the late 1970s, when the efficiency was nearly doubled to about 24 miles per gallon. Except in California, utility companies make more money when they sell more energy, so they have no incentive to conserve. Improved efficiency standards for appliances caused household energy use to decline in the U.S., until the proliferation of electronic devices that consume energy even in standby mode and a marked growth in the size of homes offset these improvements. Building-efficiency standards could have averted increased energy use, but even the existing weak standards have been difficult to enforce. High energy costs provide the most effective enforcement, because continual inspections are impractical.

Again, the solution to the climate problem requires the phasing out of carbon emissions from fossil fuels. But figure 2 shows this is not happening, because carbon capture is not being used with any of these fossil fuels. Contrary to Lovins's projection, "soft" renewable energies remain imperceptibly small. The largest carbon-free energy source is nuclear, which Lovins would eliminate. The main renewable energy source currently in use is hydroelectric, provided by large hydropower projects built in the middle of the twentieth century, which Lovins also would eliminate. The second-largest renewable energy source is biomass burning, whose "softness" is questionable. Coal, oil, and gas provide most U.S. energy. I have discussed figure 2 with Lovins, suggesting that a phasedown of fossil fuels requires a carbon tax. Lovins says that a tax is not needed.

It is no wonder that Lovins is hugely popular on the rubber-chicken circuit. But it is dangerous to listen to a siren without checking real-world data. Figure 2 shows that progress toward the all-soft-energies track has been teeny-tiny compared with what is needed. Kidding ourselves that the world will suddenly move onto the soft-energy path would sentence our grandchildren to an unhappy, deadly future.

Why do fossil fuels continue to provide most of our energy? The reason is simple. Fossil fuels are the cheapest energy. This is in part due to their marvelous energy density and the intricate energy-use infrastructure that has grown up around fossil fuels. But there is another reason: Fossil fuels are cheapest because we do not take into account their true cost to society. Effects of air and water pollution on human health are borne by the public. Damages from climate change are also falling on the public, but they will be borne especially by our children and grandchildren.

How can we fix the problem? The solution necessarily will increase the price of fossil fuel energy. We must admit that. In the end, energy efficiency and carbon-free energy can surely be made less expensive than fossil fuels, if fossil fuels' cost to society is included.

The difficult part is that we must make the transition with extraordinary speed if we are to avert climate disaster.

Rather than immediately defining a proposed framework for a solution, which may appear to be arbitrary without further information, we need to first explore the problem and its practical difficulties. Two alternative legislative actions have been proposed in the United States: "fee-and-dividend" and "cap-and-trade." Let's begin by looking at the simpler approach, fee-and-dividend. In this method, a fee is collected at the mine or port of entry for each fossil fuel (coal, oil, and gas), i.e., at its first sale in the country. The fee is uniform, a single number, in dollars per ton of carbon dioxide in the fuel. The public does not directly pay any fee or tax, but the price of the goods they buy increases in proportion to how much fossil fuel is used in their production. Fuels such as gasoline or heating oil, along with electricity made from coal, oil, or gas, are affected directly by the carbon fee, which is set to increase over time. The carbon fee will rise gradually so that the public will have time to adjust their lifestyle, choice of vehicle, home insulation, etc., so as to minimize their carbon footprint.

Under fee-and-dividend, 100 percent of the money collected from the fossil fuel companies at the mine or well is distributed uniformly to the public. Thus those who do better than average in reducing their carbon footprint will receive more in the dividend than they will pay in the added costs of the products they buy.

The fee-and-dividend approach is straightforward. It does not require a large bureaucracy. The total amount collected each month is divided equally among all legal adult residents of the country, with half shares for children, up to two children per family. This dividend is sent electronically to bank accounts, or for people without a bank account, to their debit card.

As an example, consider the point in time at which the fee will reach the level of $115 per ton of carbon dioxide. A fee of that level will increase the cost of gasoline by $1 per gallon and the average cost of electricity by around 8 cents per kilowatt-hour. Given the amount of oil, gas, and coal sold in the United States in 2007, $115 per ton will yield $670 billion. The resulting dividend will be close to $3,000 per year, or $250 per month, for each legal adult resident; a family with two or more children will receive in the range of $8,000 to 9,000 per year.

Fee-and-dividend is a progressive tax. For example, my friend Al Gore (I hope he is still my friend after this book is published) will pay a heck of a lot more than $9,000 in added costs because he owns large houses and flies around the world a lot. Given the current distribution of wealth and lifestyles, about 40 percent of people will pay more in added costs than they will get back in their dividend. For the most part, it will be those with high incomes who pay more, but not always. A poor guy who commutes a hundred miles to work every day in a clunker may pay more than he gets in his dividend (although perhaps not, if he lives in a modest-size house, doesn't do a lot of recreational motoring, and rarely takes airplane trips). Sorry, poor guy, but it is those kinds of practices that will be changed, in the long run, by a rising carbon fee. The cost will encourage the poor guy to figure out more efficient transportation or live closer to his work.

By the way, Al Gore agrees that fee-and-dividend is the best way to reduce carbon emissions, but his proposal is to reduce payroll taxes rather than give dividends to the public. I prefer the dividend because I don't trust the government to make the tax reduction balance out the fee. Also, not everybody is on a payroll. A dividend is just simpler.

Few activities would be unaffected by a carbon fee-and-dividend. Today we often import food from halfway around the world, rather than from a nearby farm, in part because there is no tax on aviation fuel. Why? Lobbying. A deal was made in the 1940s to encourage the budding aviation industry—and lobbying makes it hard to get rid of sweet deals. All sweet deals will be wiped off the books by a uniform carbon fee at the source, which will affect all fossil fuel uses.

I'm asked, "If people get a dividend, won't they just go out and spend that money on their gas-guzzler or whatever fossil fuels they have been using?" Maybe they will at first, but in the long run they will tend to adjust their decisions on vehicle choice and other matters as the carbon price gradually continues to rise.

A rising carbon price does not eliminate the need for efficiency regulations, but it makes them work much better. Building codes, for example, usually have energy efficiency requirements, but every city finds that they are impossible to enforce well. The builder changes things after inspection, or the building operation is simply inefficient. The best enforcement is carbon price—as the fuel price rises, people pay attention to waste.

Economists are almost unanimous that a uniform rising carbon fee is the least costly way to phase out fossil fuels. This allows proper competition between energy efficiency and alternative carbon-free energy sources such as solar energy, wind, and nuclear power. It also "internalizes" the incentive to reduce the use of carbon fuels, especially coal, in literally billions of decisions ranging from commuting behavior to the design of vehicles, aircraft, cities, and so forth.

"Wait a minute," you may be saying. "This carbon fee doesn't sound like the deal I have been hearing about." You are right. Most of the talk is about cap-and-trade, the basis of proposed legislation being considered by Congress, specifically Representatives Henry Waxman and Ed Markey's American Clean Energy and Security Act. Cap-and-trade is what governments and the people in alligator shoes (the lobbyists for special interests) are trying to foist on you.

Whoops. As an objective scientist I should delete such personal opinions, or at least flag them. But I am sixty-eight years old, and I am fed up with the way things are working in Washington. Foolishly, I imagined that we might really get "change" in the way things worked there. As I said, I was among those who had moist eyes on Election Day in November 2008, when President-elect Obama gave his speech in Chicago. But things are still done in the same way in Washington. No doubt I was naïve to think that it might be otherwise, and, unfortunately, so were millions of young people.

I am not blaming President Obama. On the contrary, he is still our best hope. But he must actually look into this matter, not rely on watered-down advice from his sources of information and advisers. The leaders Obama appointed in science and energy are the most knowledgeable people in the field, but there are many others in his inner circle of advisers. The stakes in the policy adopted for energy and climate are too great to be based on aggregate advice or a sum of political compromises. The present situation is analogous to that faced by Lincoln with slavery and Churchill with Nazism— the time for compromises and appeasement is over.

It is hard to blame anybody in Obama's circle of advisers, even though I detest the tactics that have infested American politics. It seems to be believed that if you don't have tough guys around you, guys who can deliver tit for tat, counterblows to attacks from the other side, maybe with similar tactics, you will soon be on the outside, looking in at somebody else governing. I don't know, maybe that is true. But I also believe that the public can appreciate a principled stand, even one that takes political hits, if it is properly explained.

The reason it is hard for me to blame Obama's advisers is that I see where they are getting their information. It is from good people, our friends, the people who are believed to be the most supportive of the environment, including climate preservation. I refer to some members of Congress who are among those with the strongest environmental voting records, such as Waxman and Markey, and I refer especially to organizations such as the Environmental Defense Fund, the Natural Resources Defense Council, and the Pew Foundation.

People tell me, "You must be wrong, because the polluters are opposed to cap-and-trade, so cap-and-trade must be good." Sure, those in the fossil fuel industry would prefer no regulations at all, so that is their first choice—they stall any action as long as possible. But they know that something is coming down the pike. And they are spending enormous amounts of money to be sure that cap-and-trade is doctored to allow as much business-as-usual emissions to continue as long as possible.

Let's discuss cap-and-trade explicitly first. Then I will provide a bottom-line proof that it cannot work. Because I have already made up my mind about the uselessness of cap-and-trade, my commentary may be slanted, but you have been warned, so you should be able to make up your own mind.

In cap-and-trade, the amount of a fossil fuel for sale is supposedly "capped." A nominal cap is defined by selling a limited number of certificates that allow a business or speculator to buy the fuel. So the fuel costs more because you must pay for the certificate and the fuel. Congress thinks this will reduce the amount of fuel you buy—which may be true, because it will cost you more. Congress likes cap-and-trade because it thinks the public will not figure out that a cap is a tax.

How does the "trade" part factor in? Well, you don't have to use the certificate; you can trade it or sell it to somebody else. There will be markets for these certificates on Wall Street and such places. And markets for derivatives. The biggest player is expected to be Goldman Sachs. Thousands of people will be employed in this trading business—the big boys, not guys working for five dollars an hour. Are you wondering who will provide their income? Three guesses and the first two don't count. Yes, it's you—sorry about that. Their profits are also added to the fuel price.

What is the advantage of cap-and-trade over fee-and-dividend, with the fee distributed to the public in equal shares? There is an advantage to cap-and-trade only for energy companies with strong lobbyists and for Congress, which would get to dole out the money collected in certificate selling, or just give away some certificates to special interests. Don't hurry to write a letter to your congressional representative asking for a certificate to pollute—that's not how things work in Washington. Your paragraph requesting a certificate is not likely to be included in the Waxman-Markey bill, even though at last count 1,400 pages had been added. Again, think lobbyists. Think revolving doors. People in alligator shoes write the paragraphs that actually get added. If you think I am kidding, ask yourself this: Do you believe that your representatives in Congress can write 1,400 pages themselves? It is still a free country, so you can hire your own lobbyist, but the price is kind of high. A coal company can afford someone like Dick Gephardt—can you?

Okay, I will try to be more specific about why cap-and-trade will be necessarily ineffectual. Most of these arguments are relevant to other nations as well as the United States.

First, Congress is pretending that the cap is not a tax, so it must try to keep the cap's impact on fuel costs small. Therefore, the impact of cap-and-trade on people's spending decisions will be small, so necessarily it will have little effect on carbon emissions. Of course that defeats the whole purpose, which is to drive out fossil fuels by raising their price, replacing them with efficiency and carbon-free energy.

The impact of cap-and-trade is made even smaller by the fact that the cap is usually not across the board at the mine. In the fee-and-dividend system, a single number, dollars per ton of carbon dioxide, is applied at the mine or port of entry. No exceptions, no freebies for anyone, all fossil fuels covered for everybody. In cap-and-trade, things are usually done in a more complicated way, which allows lobbyists and special interests to get their fingers in the pie. If the cap is not applied across the board, covering everything equally, any sector not covered will benefit from reduced fuel demand, and thus reduced fuel price. Sectors not covered then increase their fuel use.

In contrast, the fee-and-dividend approach puts a rising and substantial price on carbon. I believe that the public, if honestly informed, will accept a rise in the carbon fee rate because their monthly dividend will increase correspondingly.

Second, the cap-and-trade target level for emissions (defined by the number of permits) sets a *floor* on emissions. Emissions cannot go lower than this floor, because the price of permits on the market would crash, bringing down fossil fuel prices and again making it more economical for profit-maximizing businesses to burn fossil fuels than to employ energy-efficiency measures and renewable-energy technology. It would be akin to a drug dealer luring back former customers by offering free cash along with a free fix.

With fee-and-dividend, in contrast, we will reach a series of points at which various carbon-free energies and carbon-saving technologies are cheaper than fossil fuels plus their fee. As time goes on, fossil fuel use will collapse, remaining coal supplies will be left in the ground, and we will have arrived at a clean energy future. And that is our objective.

A perverse effect of the cap-and-trade floor is that altruistic actions become meaningless. Say that you are concerned about your grandchildren, so you decide to buy a high-efficiency little car. That will reduce your emissions but not the country's or the world's; instead it will just allow somebody else to drive a bigger SUV. Emissions will be set by the cap, not by your actions.

In contrast, the fee-and-dividend approach has no floor, so every action you take to reduce emissions helps. Indeed, your actions may also spur your neighbor to do the same. That snowballing (amplifying feedback) effect is possible with fee-and-dividend, but not with cap-and-trade.

Third, offsets cause actual emission reductions to be less than targets, because emissions covered by an offset do not count as emissions. They don't count as emissions to the politicians, but they sure count to the planet! For example, actual reductions under the Waxman-Markey bill have been estimated to be less than half of the target, because of offsets.

Fourth, Wall Street trading of emission permits and their derivatives in the anticipated multitrillion-dollar carbon market, along with the demonstrated volatility of carbon markets, creates the danger of Wall Street failures and taxpayer-funded bailouts. In the best case, if market failures are avoided, there is the added cost of the Wall Street trading operation and the profits of insider trading. To believe that there will be no insider profits is to believe that government overseers are more clever than all the people on Wall Street and that there is no revolving door between Wall Street and Washington. Where will Wall Street profits come from? They too will come from John Q. Public via higher energy prices.

In contrast, a simple flat fee at the mine or well, with simple long division to determine the size of the monthly dividend to all legal residents, provides no role for Wall Street. Could that be the main reason that Washington so adamantly prefers cap-and-trade?

Fee-and-dividend is revenue neutral to the public, on average. Cap-and-trade is not, because we, the public, provide the profits to Wall Street and any special interests that have managed to get written into the legislation. Of course Congress will say, "We will keep the cost very low, so you will hardly notice it." The problem is, if it's too small for you to notice, then it is not having an effect. But maybe Congress doesn't really care about your grandchildren.

Hold on! Or so you must be thinking. If cap-and-trade is so bad, why do environmental organizations such as the Environmental Defense Fund and the National Resources Defense Council support it? And what about Waxman and Markey, two of the strongest supporters of the environment among all members of the House of Representatives?

I don't doubt the motives of these people and organizations, but they have been around Washington a long time. They think they can handle this problem the way they always have, by wheeling and dealing. Environmental organizations "help" Congress in the legislative process, just as the coal and oil lobbyists do. So there are lots of "good" items in the 1,400 pages of the Waxman-Markey bill, such as support for specific renewable energies. There may be more good items than bad ones—but unfortunately the net result is ineffectual change. Indeed, the bill throws money to the polluters, propping up the coal industry with tens of billions of taxpayer dollars and locking in coal emissions for decades at great expense.

Yet these organizations say, "It is a start. We will get better leg-islation in the future." It would surely require continued efforts for many decades, but we do not have many decades to straighten out the mess.

The beauty of the fee-and-dividend approach is that the carbon fee helps any carbon-free energy source, but it does not specify these sources; it lets the consumer choose. It does not cost the gov-ernment anything. Whether it costs citizens, and how much, de-pends on how well they reduce their carbon footprint.

A quantitative comparison of fee-and-dividend and cap-and-trade has been made by economist Charles Komanoff (www.komanoff .net/fossil/CTC_Carbon_Tax_Model.xls). If the carbon fee increases by $12.50 per ton per year, Komanoff estimates that U.S. carbon emissions in 2020 would be 28 percent lower than today. And that is without the snowballing (amplifying feedback) effect I men-tioned above. By that time the fee would add just over a dollar to the price of a gallon of gasoline, but the reduction in fossil fuel use would tend to reduce the price of raw crude. The 28 percent emis-sions reduction compares with the Waxman-Markey bill goal of 17 percent—which is, however, fictitious because of offsets. This approach, small annual increases of the carbon fee (ten to fifteen dollars per ton per year), is essentially the bill proposed by Con-gressman John B. Larson, a Democrat in the U.S. House of Repre-sentatives. Except Larson proposes using the money from the fee to reduce payroll taxes, rather than to pay a dividend to legal residents. The Democratic leadership and President Obama, so far, have cho-sen to ignore Congressman Larson.

A final comment on cap-and-trade versus fee-and-dividend. Say an exogenous development occurs, for example, someone invents an inexpensive solar cell or an algae biofuel that works wonders. Any such invention will add to the 28 percent emissions reduction in the fee-and-dividend approach. But the 17 percent reduction un-der cap-and-trade will be unaffected, because the cap is a floor. Per-mit prices would fall, so energy prices would fall, but emission reductions would not go below the floor. Cap-and-trade is not a smart approach.

But, you may ask, was it not proven with the acid rain problem that cap-and-trade did a wonderful job of reducing emissions at low cost? No, sorry, that is a myth—and worse. In fact, examination of the story about acid rain and power plant emissions shows the dangers in both horse-trading with polluters and the cap-and-trade floor.

Here is essentially how the acid rain "solution" worked. Acid rain was caused mainly by sulfur in coal burned at power plants. A cap was placed on sulfur emissions, and power plants had to buy permits to emit sulfur. Initially the permit price was high, so many utilities decided to stop burning high-sulfur coal and to replace it with low-sulfur coal from Wyoming. From 1990 to today, sulfur emissions have been cut in half. A smaller part of the reduction was from the addition of sulfur scrubbers to some power plants that could install them for less than the price of the sulfur permits, but the main solution was use of low-sulfur coal. Now what the dickens does that prove?

It proves that in a case where there are a finite number of point sources, and there are simple ways to reduce the emissions, and you are satisfied to just reduce the emissions by some specified fraction, then emission permits make sense. The utilities that were closest to the Wyoming coal or that needed to install scrubbers for other reasons could reduce their emissions, and so overall the cost of achieving the specified reduction of sulfur emissions was minimized. But the floor of this cap-and-trade approach prevented further reductions. Analyses have shown that the economic benefits of further reductions would have exceeded costs by a factor of twenty-five. So, in some sense, the acid rain cap-and-trade solution was an abject failure.

It is worse than that. The horse-trading that made coal companies and utilities willing to allow this cap-and-trade solution did enormous long-term damage. (What do I mean by "coal companies and utilities *willing* to allow"? That is the way it works in Washington. Special interests have so much power, or Congress chooses to give them so much sway, that their assent is needed.) The horse-trading was done in 1970. Senator Edmund Muskie, one of the best friends that the environment has ever had, felt it was necessary to compromise with the coal companies and utilities when the 1970 Clean Air Act was defined. So he allowed old coal-fired power plants to be "grandfathered": they would be allowed to continue to pollute, because they would soon be retired anyhow, or so the utilities said. Like fun they would. Those old plants became cash cows once they were off the pollution hook—the business community will never let them die. Thousands of environmentalists have been fighting those plants and trying to adjust clean air regulations ever since. Yet today, in 2009, there are still 145 operating coal-fired power plants in the United States that were constructed before 1950. Two thirds of the coal fleet was constructed before the Clean Air Act of 1970 was passed.

Those people, including the leaders of our nation, who tell you that the acid rain experience shows that cap-and-trade will work for the climate problem do not know what they are talking about. The experience with coal-fired power plants does contain important lessons, though.

First, it shows that the path we start on is all-important. People who say that cap-and-trade is a good start and we will move on from there are not looking at reality. Four decades later we are still paying for an early misstep with coal-fired plants.

Second, it shows that we need a simple, across-the-board solution that covers all emissions. A fee or tax must be applied at the source. If Congress insists that it must help somebody who will be hurt by the carbon fee, such as coal miners, fine—Congress can provide for job retraining or some other compensation. But the fee on fossil fuel carbon must be uniform at the source, with no exceptions.

Finally, let me address the ultimate defense that is used for cap-and-trade: "The train has left the station. It is too late to change. President Obama has decided. The world has decided. It must be cap-and-trade, because an approach such as you are talking about would delay things too much." That latter claim turns truth on its head, calling black "white" and white "black." The truth is shown by empirical evidence. In February 2008, British Columbia decided to adopt a carbon tax with an equal reduction of payroll taxes. Five months later it was in place and working. This year there was an election in British Columbia in which the opposition party campaigned hard against the carbon tax. They lost. The public liked the carbon tax with a payroll tax reduction. Now both parties support it. In contrast, it took a decade to negotiate the cap-and-trade Kyoto Protocol, and many countries had to be individually bribed with concessions. The result: slow implementation and an ineffectual reduction of emissions. The Waxman-Markey bill is following a similar path.

I almost forgot that I had agreed to provide a proof that the approach pursued by governments today cannot conceivably yield their promise of an 80 percent emission reduction by 2050. It is an easy proof. An 80 percent reduction in 2050 is just what occurs if coal emissions are phased out between 2010 and 2030, as shown in figure 26. This is based on the moderate oil and gas reserves estimated by IPCC—implying also that we cannot go after the last drops of oil. First ask if governments are building any new coal plants. The answer: "Lots of them." Then ask how they will persuade the major oil-producing nations to leave their oil in the ground. The answer: "Duh." Proof complete.

Okay, at long last, we can address the fundamental problem. What is the backbone and framework for a solution to human-caused climate change?

The backbone must be a rising fee (tax) on carbon-based fuels, uniform across the board. No exceptions. The money must be returned to the public in a way that is direct, so they realize and trust that (averaged over the public) the money is being returned in full. Otherwise the rate will never be high enough to do the job. Returning the money to the public is the hard part in the United States. Congress prefers to keep the money for itself and divvy it out to special interests.

The framework concerns how to make an across-the-board fee on fossil fuel carbon work on a global basis, in a way that is fair, because unless there is a universal carbon fee, it will be ineffective. The backbone, I will argue, makes it relatively simple to define international arrangements—I will explain what I mean by "relatively simple" in a moment. The backbone also makes it practical to have a framework that deals with the problem of fairness between those who have caused the problem, those who are causing the problem, and those who are primarily the victims of others. The framework can also help deal with the fundamental problems of population and poverty.

Contrary to the assertion by proponents of a Kyoto-style cap-and-trade agreement, cap-and-trade is not the fastest way to an international agreement. That assertion is another case of calling black "white," apparently under the assumption that the listener will accept it without thinking. A cap-and-trade agreement will be just as hard to achieve as was the Kyoto Protocol. Indeed, why should China, India, and the rest of the developing world accept a cap when their per-capita emissions are an order of magnitude less

than America's or Europe's? Leaders of developing countries are
making that argument more and more vocally. Even if differences
are papered over to achieve a cap-and-trade agreement at upcoming
international talks, the agreement is guaranteed to be ineffectual.
So eventually (quickly, I hope!) it must be replaced with a more
meaningful approach. Let's define one.

The key requirement is that the United States and China agree to
apply across-the-board fees to carbon-based fuels. Why would China
do that? Lots of reasons. China is developing rapidly and it does not
want to be saddled with the fossil fuel addiction that plagues the
United States. Besides, China would be hit at least as hard as the
United States by climate change. The most economically efficient
way for China to limit its fossil fuel dependence, to encourage energy efficiency and carbon-free energies, is via a uniform carbon fee.
The same is true for the United States. Indeed, if the United States
does not take such an approach, but rather continues to throw life-
lines to special interests, its economic power and standard of living
will deteriorate, because such actions make the United States economy less and less efficient relative to the rest of the world.

Agreement between the United States and China comes down to
negotiating the ratio of their respective carbon tax rates. In this negotiation the question of fairness will come up—the United States
being more responsible for the excess carbon dioxide in the air today despite its smaller population. That negotiation will not be
easy, but once both countries realize they are in the same boat and
will sink or survive together, an agreement should be possible.

Europe, Japan, and most developed countries would likely agree
to a status similar to that of the United States. It would not be difficult to deal with any country that refuses to levy a comparable
across-the-board carbon fee. An import duty could be collected by
countries importing products from any nation that does not levy
such a carbon fee. The World Trade Organization already has rules
permitting such duties. The duty would be based on standard estimates of the amount of fossil fuels that go into producing the im-
ported product, with the exporting company allowed the option of
demonstrating that its product is made without fossil fuels, or with
a lesser amount of them. In fact, exporting countries would have a
strong incentive to impose their own carbon fee, so that they could
keep the revenue themselves.

As for developing nations, and the poorest nations in the world, how can they be treated fairly? They also must have a fee on their fossil fuel use or a duty applied to the products that they export. That is the only way that fossil fuels can be phased out. If these countries do not have a tax on fossil fuels, then industry will move there, as it has moved already from the West to China and India, with carbon pollution moving along with it. Fairness can be achieved by using the funds from export duties, which are likely to greatly exceed foreign aid, to improve the economic and social well-being of the developing nations.

I do not want to wander far into these subjects, but it would be inappropriate not to mention the connection between population and climate change. The stress that humans place on the planet and other species on the planet is closely related to human population growth. Stabilization of atmospheric composition and climate almost surely requires a stabilization of human population.

The encouraging news is that there is a strong correlation between reduced fertility rates, increased economic well-being, and women's rights and education. Many Western countries now have fertility rates below or not far from the replenishment level. The substantial funds that will necessarily be generated by an increasing fee on fossil fuel carbon should be used to reward the places that encourage practices and rights that correlate with sustainable populations.

In summary, the backbone of a solution to the climate problem is a flat carbon emissions price applied across all fossil fuels at the source. This carbon price (fee, tax) must rise continually, at a rate that is economically sound. The funds must be distributed back to the citizens (not to special interests)—otherwise the tax rate will never be high enough to lead to a clean energy future. If your government comes back and tells you that it is going to have a "goal" or "target" for carbon emission reductions, even a "mandatory" one, you know that it is lying to you, and that it doesn't give a damn about your children or grandchildren. For the moment, let's assume that our governments will see the light.

Once the necessity of a backbone flat carbon price across all fossil fuel sources is recognized, the required elements for a framework agreement become clear. The principal requirement will be to define how this tax rate will vary between nations. Recalcitrance of any nations to agree to the carbon price can be handled via import duties, which are permissible under existing international agreements. The framework must also define how proceeds of carbon duties will be used to assure fairness, encourage practices that improve women's rights and education, and help control population. A procedure should be defined for a regular adjustment of funds' distribution for fairness and to reward best performance.

Well, what happens if, instead of accepting the need for a rising carbon price, our governments continue to deceive us, setting goals and targets for carbon emissions reductions?

In that case we had better start thinking about the Venus syndrome.

CHAPTER 10

The Venus Syndrome

IN DECEMBER 2008 I HAD THE HONOR of giving the Bjerknes Lecture, a one-hour talk at the annual meeting of the American Geophysical Union in San Francisco, named for Vilhelm Bjerknes, the Norwegian physicist and meteorologist who was a founding father of modern weather forecasting. My talk was titled "Climate Threat to the Planet."

I realized that, in addition to reviewing current understanding of ongoing climate change, I had better include a look at Earth from a planetary perspective. "A planet in peril" had become a popular phrase, but it seemed that people using it did not understand the full implications.

Also, I was beginning to question a basic presumption contained in my first comprehensive paper, "Climate Impact of Increasing Atmospheric Carbon Dioxide," published in *Science* in 1981. That paper showed the dominant role that coal would have in future climate change and predicted that global climate change would rise above the level of natural climate variability by the end of the twentieth century. My presumption was that, as the reality of climate change became apparent, government policies would begin to be adapted in a rational way. Since then, two trends had become clear, suggesting that my presumption could be disastrously wrong.

First, special interests were remarkably successful in preventing the public at large from understanding the situation. The result was a growing gap between what was understood by the relevant scientific community about human-caused climate change and what was appreciated by the public.

Second, it had become clear that greenwash was a near universal response of politicians to the climate change issue. I became well acquainted with greenwash via interactions with several governors, as summarized in the "Dear Governor Greenwash" letter on my Web site, and I observed that the media allowed politicians to get away with what amounted to fake environmentalism. Most important, it was becoming apparent that the international follow-up to the ineffectual Kyoto Protocol could be another ineffectual target-based cap-and-trade agreement. And several nations, including the United States, seemed to be going right ahead with plans for coal-to-liquid fuels and development of unconventional fossil fuels, oblivious to the long-term implications.

What can one do in such a situation? Writing scientific papers, giving talks, writing op-eds did not seem to have any effect in Washington or other capitals. There are thousands of oil, gas, and coal lobbyists in Washington. These lobbyists are very well paid. It is no wonder that government energy policies are so hospitable to the fossil fuel industry.

Given this situation, it seems possible that strategic changes to fossil fuel use will not be adopted. Goals, cap-and-trade, and offsets—in other words, business as usual—may continue. So we had better examine what may happen if we push the planet beyond its tipping point.

I began my lecture with a discussion of the Venus syndrome, showing the "Goldilocks" chart (figure 29) I had used in my Iowa talk in 2004. Earth is the only one of the three terrestrial planets that is "just right" for life to exist. Mars is too cold. Venus is too hot. The temperatures of these planets are affected by the distance of each planet from the sun and by the planet's albedo, the fraction of sunlight it reflects to space. But their surface temperatures are also strongly influenced by the amount of atmospheric greenhouse gases.

Mars has so little gas in its atmosphere that its greenhouse effect is negligible, and the surface temperature averages about –50 degrees Celsius (about 60 degrees below zero Fahrenheit). Greenhouse gases warm Earth's surface by about 33 degrees Celsius (about 60 degrees Fahrenheit), making the average surface temperature about 15 degrees Celsius (about 60 degrees Fahrenheit). Venus has so much carbon dioxide in its atmosphere that it has a greenhouse warming of

−50°C   +15°C   +450°C

FIGURE 29. *Earth is the "Goldilocks" planet, not too hot, not too cold, just right for life to exist.*

several hundred degrees, with the surface, at 450 degrees Celsius (about 850 degrees Fahrenheit), hot enough to melt lead.

Venus is almost as big as Earth, with a diameter about 95 percent as large. Venus and Earth, having condensed from the same interstellar gas and dust during the formation of the solar system, must have begun with similar atmospheric compositions. So the early Venus atmosphere contained lots of water vapor. The sun was 30 percent dimmer at that time, so Venus was probably cool enough to have oceans on its surface. But they did not last long. As the sun brightened, the surface of Venus became hotter, water evaporated, and the strong greenhouse effect of water vapor amplified the warming. Eventually a "runaway" greenhouse effect occurred, with the ocean boiling or evaporating into the atmosphere. The surface became so hot that all the carbon dioxide in the crust was "baked out" into the atmosphere. There was a lot of carbon in the crust, so much that the atmosphere became predominately carbon dioxide. The atmosphere of Venus is now almost 97 percent carbon dioxide and the surface pressure on Venus is 90 bars, i.e., 90 times greater than the surface pressure on Earth—that's about 1,300 pounds per square inch, which would crush any human visitors, if they were not fried first.

The water vapor on Venus was eventually lost to space. Ultraviolet sunlight "dissociates" (breaks up) atmospheric water vapor molecules into hydrogen and oxygen. The molecules and atoms are continuously moving about in the atmosphere. After dissociation, some light hydrogen atoms are able to escape the planet's gravitational field. The remaining oxygen combines with other material, for example, with carbon, to make carbon dioxide. In this way, water was lost from Venus.

Can we confirm this explanation for why Venus has no water today, while it must have had water at the time of its origin? Yes. We have measurements of hydrogen isotopes in the Venus atmosphere. Deuterium, which is heavy hydrogen with a nucleus containing a neutron in addition to the usual proton, is ten times more abundant on Venus relative to normal hydrogen than it is on Earth and on the sun, even though Venus, Earth, and the sun all formed from the same primordial nebula. The enrichment of heavy hydrogen on Venus provides a measure of its lost hydrogen, because the lighter, normal hydrogen can escape the planet's gravitational field more easily than heavier deuterium can escape. The data agree with the assumption that an early Venus was wet.

So Venus had a runaway greenhouse effect. Could Earth? Of course we know that it could. The question is, rather, how much must carbon dioxide (or some other climate forcing) increase before a runaway effect occurs?

One way to address that question is with climate models. I have mentioned that we need to treat climate models with skepticism, but if we recognize their assumptions and limitations, and find ways to test them against reality, they can aid our analysis. In my Bjerknes lecture, I showed the graph in figure 30, which was taken from the "Efficacy of Climate Forcings" paper I published with several coauthors in 2005. This graph is the calculated global temperature change at the end of a hundred-year climate simulation divided by the climate forcing. In other words, it shows how sensitive the climate is to either a negative (left side of the graph) or a positive (right side) climate forcing. But it shows only a partial response because of the brevity of the simulation and the exclusion of slow feedbacks such as ice sheet change. Figure 30 illustrates results of experiments with two different climate forcings: changing atmospheric carbon dioxide and changing brightness of the sun.

The climate sensitivity of the model began to increase rapidly with either a large negative forcing or a large positive forcing. Qualitatively, this is the behavior that we know must occur: A sufficient negative forcing causes a runaway snowball Earth condition, with freezing temperatures over the entire planet, while a sufficient positive forcing causes a runaway greenhouse effect. We know that this

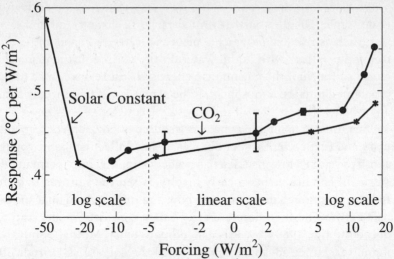

FIGURE 30. *Global temperature change in a climate model per unit forcing. Data from Hansen et al., "Efficacy of Climate Forcings." See sources for chapter 1.)*

U-shaped curve is correct—the question is, at what forcings do the sharp upturns to runaway conditions occur?

There may have been problems with the model, inaccuracies in the representation of climate processes, which would have caused the upswings in sensitivity to occur at too small forcings. However, the largest uncertainties that we can identify work in the opposite direction. The model employed to calculate figure 30 had fixed ice sheet area. If ice sheets had been allowed to grow with negative forcing and melt with positive forcing, and if enough time had been allowed for the melting to occur, the sharp upturns would have taken place at smaller forcings. Also, limited empirical evidence suggests that as the planet gets warmer, the amount of other trace greenhouse gases in the air, in addition to carbon dioxide and water vapor, tends to increase.

These considerations, and the modeling result, suggest that the forcings needed to reach snowball Earth or runaway greenhouse conditions are no more than 10 to 20 watts per square meter when solar irradiance or carbon dioxide change are defined as the forcing. The change required for a snowball Earth is of course a negative (reduced) forcing.

We can state this result in another way that may be easier to understand. There is only a limited range of distance around any star, such as the sun, at which a planet will have a "habitable" surface temperature, with liquid water on the surface. If the planet is closer to the sun, the greenhouse effect will cause any water to be evaporated into the atmosphere. If the planet is too far from the sun, any water will be frozen all the way to the equator.

This limited habitable zone was a source of puzzlement to planetary and Earth scientists for decades. It was called the "faint young sun" paradox. How had Earth avoided slipping into permanent snowball conditions in its early history, when the sun was known to have been much dimmer? And how had life survived on a snowball Earth? The first simple energy-balance climate model, introduced in the 1960s by Russian climatologist Mikhail Budyko, found that, if ice advanced as far toward the equator as latitude 30 degrees, the amplifying feedback of increased planetary albedo would cause ice to advance suddenly all the way to the equator. The resulting ice-covered Earth would reflect most sunlight, so its climate should be stable in this snowball condition, even if the sun's brightness increased as much as several percent.

A solution to the paradox became clear in the 1990s, by which time geologic evidence for Earth's history was quite detailed. In fact, Earth *had* fallen into the snowball state, several times, with ice extending all the way to the equator. The flaw in the 1960s thinking was the assumption that Earth could not emerge from the snowball. The explanation, suggested by Joseph Kirschvink in 1992, and investigated in greater depth by Paul Hoffman and Daniel Schrag, was that the weathering process that takes carbon dioxide out of the air would cease on a snowball Earth. But continental drift and volcanic eruptions would continue. Therefore carbon dioxide would build up in the atmosphere until there was a strong enough greenhouse effect to begin to melt ice at the equator. At that point, the amplifying feedback of darker ocean replacing ice caused rapid ice melt and further global warming.

Another flaw in the 1960s thinking was caused by the simplicity of Budyko's energy balance calculation. It turns out that a realistic three-dimensional climate model, including ocean dynamics and seasonal and daily variations of sunlight, does not yield a hard iceball, with the ocean covered everywhere by a thick solid ice layer.

Indeed, snowball Earth was more like a slushball. The areas of open water make it easier to understand how life survived the snowball state.

Life seemed to be influenced profoundly by the final snowball event, which occurred about 600 million years ago. Prior to that time, the most complex organisms on Earth were unicellular protozoa and filamentous algae—in other words, the only life on the planet was green scum. The final snowball was followed promptly by the Cambrian explosion of life. Eukaryotes, cells with a membrane-bound nucleus, expanded rapidly into eleven different body plans. These eleven animal phyla still encompass all animals that have ever inhabited Earth.

At the end of the last snowball Earth, the sun's brightness was within 6 percent of its present value. There will never be another snowball Earth, because the sun continues to get brighter. In fact, with humans on the planet, there will never be another ice age. Sorry to distract you with an aside, but I need to clarify a point, one that is relevant to the present discussion.

A few geologists continue to speak as if they expect Earth to proceed into the next glacial cycle, just as it would have if humans were not around. That glacial period would begin with an ice sheet developing and growing in northern Canada. But why would we allow such an ice sheet to grow, and flow, and eventually crush major cities, when we could prevent it with the greenhouse gases from a single chlorofluorocarbon factory? Humans are now "in charge" of future climate. It is a trivial task to avoid the negative net climate forcing that would push the planet into an ice age (moving conditions toward the left in figure 30). But it is not an easy task to find a way to stop the growth of atmospheric greenhouse gases, most notably carbon dioxide (which moves conditions toward the right in figure 30), as we have been discussing.

How will the sun's continuing evolution alter Earth's climate on long time scales? Our sun is a very ordinary medium-size star. It is about 4.6 billion years old, still "burning" hydrogen, producing helium by nuclear fusion in the sun's core, releasing energy in the process. It is slowly getting brighter. As the hydrogen fuel is exhausted, leaving inert helium in the core, the sun will expand enormously to its Red Giant phase as it burns hydrogen in its outer shell. The expanding sun will toast and eventually swallow Earth about 5 billion years from now. Nothing for you or your grandchildren to worry about. By that time, if humanity still exists, the people 200 million generations from now may have the technology to escape to another solar system.

Humanity will need to figure out climate control long before our sun approaches the Red Giant phase. In one billion years the sun will be about 10 percent brighter than it is today. That climate forcing, about 25 watts per square meter, is surely enough to push Earth into the runaway greenhouse effect, evaporate the oceans, and exterminate all life on the planet. But you and your grandchildren do not need to worry about the long-term change of the sun's brightness either, because that trend is negligible in comparison with what humans are doing by adding greenhouse gases to the air. Besides, long before the sun becomes 10 percent brighter, humans will have realized that they need to shade the sun a bit, if they want life of the sort that we know to continue. The required "geo-engineering" would be of a simple, direct kind, reflecting a fraction of incident sunlight back to space, a task that will surely be easy for a civilization that exists in future millennia, assuming that it exists.

This geo-engineering comment requires one more digression, to answer the inevitable question: Why not use such a geo-engineering trick to solve our present global warming problem, thus avoiding the need to draw down carbon dioxide to less than 350 parts per million? There are several reasons. First, carbon dioxide must be less than 350 ppm to avoid ocean acidification problems. Second, sun shielding at present is far more expensive and difficult to implement than rational alternatives such as energy efficiency, renewable energy, and nuclear power. Third, it is generally a bad idea to try to cover up one pollution effect by introducing another; such an approach is likely to have many unintended effects. It is hard to match nature. Better to keep atmospheric composition and solar irradiance at the levels to which humanity and nature are adapted. The purpose of sun shielding in the very distant future would be to keep solar irradiance at the level to which life is adapted.

Allow me to elaborate just a bit on the second of these reasons, why implementation of such geo-engineering does not make sense now. Geo-engineering costs money. In contrast, some of the more attractive alternatives would more than pay for themselves. Pay-for-itself is true, for example, for energy efficiency and nuclear power, at least in the mode that nuclear power would be used in places such as India and China, countries that would be expected to choose modular designs and limit the ability of antinukes and bureaucratic lethargy to delay construction and drive up costs. The earliest third- and fourth-generation nuclear power plants will be expensive relative to coal without carbon capture, but nations that choose to limit construction delays should be able to produce nearly carbon-free nuclear energy that is cost-effective. Some renewable energies are expensive relative to fossil fuels, but there are instances where renewable energy is already cost-effective, and these instances should increase with future economies of scale. Although first priority should be given to energy efficiency, renewable energies, and nuclear power, it does make sense to carry out geo-engineering research to define options in the event that continued business-as-usual energy policies create a planetary emergency that demands rapid changes.

Now we are ready for the important part—trying to figure out how close we are to the climate forcing that will cause a runaway greenhouse effect. Until recently I did not worry much about that. Why? Because I knew that at some times in the past there was much more carbon dioxide in the air than today, probably a few thousand parts per million. Even burning all of the fossil fuels will not exceed that amount, so we should be safe, right?

Wrong, unfortunately. It turns out that there are three factors or circumstances that alter the picture, and each of them works in the bad direction.

Circumstance 1 is not the biggest factor, but I start with it because it is substantial and we understand it accurately. Circumstance 1 is the irradiance of the sun. At earlier times, when atmospheric carbon dioxide was more abundant, the sun was dimmer. For example, 250 million years ago the sun was about 2 percent dimmer than it is now. A 2 percent change of solar irradiance is equivalent to doubling the amount of carbon dioxide in the atmosphere. So if the estimated amount of carbon dioxide 250 million years ago was 2,000 ppm, it would take only about 1,000 ppm of carbon dioxide today to create a climate equally as warm, assuming other factors are equal. (As explained earlier, a 2 percent change of solar irradiance and a doubling of carbon dioxide are equivalent forcings, each being about 4 watts per square meter.) In other words, the fact that some scientists have estimated that CO_2 was much larger earlier in Earth's history, perhaps even by a few thousand parts per million, does not mean that we could tolerate that much carbon dioxide now without hitting runaway conditions, because the sun is brighter now.

Circumstance 2 is the "measurement," or estimation, of past carbon dioxide amounts. Actually, we have direct measurements of past carbon dioxide only for recent glacial to interglacial climate changes, the period with ice core data—essentially the past million years. Maximum carbon dioxide amount in that period was about 300 ppm, until humans started burning fossil fuels. Estimates for more ancient times are based on indirect ("proxy") measures, but such indirect inferences have great uncertainties. Results from many different methods are compared in our 2008 "Target Atmospheric CO_2" paper's supplementary material. Some methods yield carbon dioxide amounts in the early Cenozoic era, between 65 and 50 million years ago, as great as 2,000 ppm, but other methods suggest that the maximum amount was less than 1,000 ppm.

Certain methods of estimating ancient carbon dioxide levels explicitly depend on assumptions about how much carbon dioxide would have been needed to cause the recorded climate change. In other words, these methods depend on assumed climate sensitivity. For example, with standard assumptions about climate sensitivity, it was estimated that unfreezing a hard "iceball" Earth—an Earth with oceans frozen solid to depths of one kilometer or more—would require a huge amount of carbon dioxide. However, we now recognize that a hard iceball is an unrealistic picture of what actually would have been snowball Earth conditions. Also, the transient phase of unfreezing snowball Earth is not directly relevant to figuring out the climate forcing needed for the runaway greenhouse effect. In other words, although the carbon dioxide amount in the air may have been large just before and during the planet's thawing, atmospheric carbon dioxide would decrease dramatically during the thawing process, long before the planet could reach runaway greenhouse conditions.

The Cenozoic era is the best period for obtaining an empirical evaluation of how near Earth may be to runaway greenhouse conditions. It provides more accurate data than earlier times, yet it encompasses much warmer climates than today, including an ice-free planet. Moreover, the Cenozoic includes the Paleocene Eocene thermal maximum (PETM), the rapid warming event that is especially relevant to our planet's future. Indeed, a new analysis of the PETM by Richard Zeebe, James Zachos, and Gerald Dickens, published in *Nature Geoscience* in mid-2009, contains, I believe, profound implications for life on the planet.

To understand the significance of the new PETM analysis and its relevance to the runaway greenhouse effect, we need to go back to figure 18 (page 153), which shows the deep ocean temperature over the past 65 million years. In the "Target Atmospheric CO_2" paper, we used that temperature curve to estimate the carbon dioxide amount over the 65-million-year period, given a very simple assumption about climate sensitivity. Specifically, we assumed that the "fast feedback" climate sensitivity was 3 degrees Celsius (for doubled carbon dioxide) for the entire 65 million years. Total climate sensitivity was greater during the most recent 34 million years because of slow feedbacks, specifically changes of ice sheet area. In justifying the assumption of a 3-degree sensitivity for the ice-free planet, we pointed to the fact that today's climate seems to be in the middle of a rather flat portion of the (fast feedback) climate sensitivity curve shown in figure 30. We also felt that it was best to employ the simplest assumption until better information became available. Under the assumption of a 3-degree Celsius climate sensitivity, we inferred that the maximum carbon dioxide amount was probably in the range of 1,000 to 1,400 ppm.

Figure 30 suggests that warmer climates may have larger climate sensitivity, indeed, that today's climate is not terribly far from the runaway situation. However, it is a model result—other models may differ. Empirical results are more meaningful. The Zeebe-Zachos-Dickens paper provides such an empirical result. In it, they show, based on the depths to which the ocean acidified and dissolved carbonate sediments, that the carbon increase that caused the PETM warming was at most 3,000 gigatons of carbon. They infer that atmospheric carbon dioxide would have increased about 700 ppm, from a baseline of approximately 1,000 ppm to about 1,700 ppm. Such a carbon dioxide increase, less than a doubling, would increase global temperature about 2 degrees Celsius, if doubled carbon dioxide sensitivity is 3 degrees Celsius. Zeebe, Zachos, and Dickens conclude, "Our results imply a fundamental gap in our understanding of the amplitude of global warming associated with large and abrupt climate perturbations."

Their conclusion is the appropriate one from a scientific perspective. I believe we can take it one step further, suggesting that their analysis is evidence that climate sensitivity in the warmer early Cenozoic was greater than 3 degrees Celsius for doubled carbon dioxide. It also favors a smaller value for the carbon dioxide abundance in the early Cenozoic. Both inferences (that the carbon dioxide level in the early Cenozoic may have been less than generally assumed, and that climate sensitivity in the early Cenozoic was greater than today) are reason for increased concern about the long-term effects of burning all fossil fuels. The PETM results would be easier to understand if the baseline carbon dioxide, prior to the PETM warming, was closer to 500 ppm. But even so, the magnitude of the PETM warming implies a climate sensitivity greater than 3 degrees Celsius for doubled carbon dioxide.

My conclusion regarding Circumstance 2 is that recent data suggest that past carbon dioxide amounts were not as great as once believed. These empirical paleoclimate data also suggest that climate sensitivity was greater when the planet was warmer, consistent with the world having been closer to the runaway greenhouse conditions when the carbon dioxide amount was greater.

Circumstance 3 concerns the time scale of climate forcings and response. Carbon dioxide that caused climate change during Earth's history was introduced much more slowly than the human-made perturbation. Slower introduction allows negative (diminishing) feedbacks in the carbon cycle to come into play. Even the solid Earth reservoir takes up carbon on millennial time scales. The negative feedbacks are the reason that, after the rapid injection of methane (which is quickly oxidized to form carbon dioxide) during the PETM, the carbon dioxide amount and global temperature recovered on fairly rapid geologic time scales (figure 18 on page 153).

The human injection of fossil fuel carbon into the atmosphere, if we choose to burn all fossil fuels, will occur so fast, on the time scale of a century or two, that carbon cycle diminishing feedbacks will not have time to come into play. If we burn all fossil fuels, the forcing will be at least comparable to that of the PETM, but it will have been introduced at least ten times faster. The time required for the ocean to respond to this forcing is only centuries. Thus, carbon cycle diminishing feedbacks will not significantly reduce the ocean warming. The warming ocean can be expected to affect methane hydrate stability at a rate that could exceed that in the PETM, where the rate of change was driven by the speed of the methane hydrate climate feedback, not by the nearly instantaneous introduction of all fossil fuel carbon.

Allow me to briefly review a few facts about the PETM, which we covered in chapter 8. Numerous studies suggest that PETM warming of 5 to 9 degree Celsius was caused by the injection of an estimated 3,000 gigatons of carbon (3,000 billion tons of carbon), although some estimates of the carbon injection were only about half that large. The Zeebe-Zachos-Dickens paper increases confidence that the PETM carbon injection did not exceed 3,000 gigatons and it draws attention to the inconsistency of such a (moderate!) carbon injection with the 5- to 9-degree-Celsius global warming, if climate sensitivity at that time were only 3 degrees Celsius for doubled carbon dioxide. Three thousand gigatons is approximately the amount of carbon contained in the sum of oil, gas, and coal fossil fuels today. However, PETM carbon could not have been from the fossil fuels, as there was no plausible mechanism for the unearthing and burning of all of the fossil fuels at that time. Indeed, it can be inferred from the carbon's isotopic signature, as explained in chapter 8, that the PETM injection was caused by the melting of methane hydrates. There were dramatic changes in ocean circulation at the time of the PETM, with deep water formation shifting from the southern hemisphere around Antarctica to the northern hemisphere. It seems probable that the warmer deep water accompanying this circulation change initiated the methane hydrate destabilization.

The time scale for the ocean temperature to largely respond to a forcing, by itself, is only centuries. But if humans burn all fossil fuels, the ice sheets will begin to disintegrate, cooling the high-latitude oceans temporarily, and delaying full climate response to the forcing. The high-latitude cooling will have important consequences in the twenty-first century, to be discussed in chapter 11. However, the cooling effect of icebergs will not significantly increase the time needed for the global ocean to warm in response to a burning of all fossil fuels. It requires less than 10 watt-years of energy, averaged over the planet, to melt enough glacial ice to raise sea level one meter and increase the meltwater temperature to the global average ocean surface temperature. Once ice sheets begin to disintegrate rapidly, the planetary energy imbalance is likely to reach several watts. So even if the entire volume of ice on the planet, equivalent to about 75 meters (almost 250 feet) of sea level, were disgorged to the ocean, the planetary energy imbalance would provide enough energy to melt all of the ice within a century or so.

My conclusion regarding Circumstance 3, the time scales, is that they would largely work against us if we were to burn all fossil fuels. Carbon cycle diminishing feedbacks, which were important for keeping Earth away from runaway conditions during paleoclimate global warming events, are not likely to be as effective in drawing down atmospheric carbon dioxide during the very rapid burning of fossil fuels by humanity. Ocean thermal inertia slows global warming, allowing more greenhouse gas to accumulate before the public takes notice of climate change, but most of the climate response to fossil fuel emissions will occur within centuries, much of it within the lifetimes of our children and grandchildren.

The paleoclimate record does not provide a case with a climate forcing of the magnitude and speed that will occur if fossil fuels are all burned. Models are nowhere near the stage at which they can predict reliably when major ice sheet disintegration will begin. Nor can we say how close we are to methane hydrate instability. But these are questions of when, not if. If we burn all the fossil fuels, the ice sheets almost surely will melt entirely, with the final sea level rise about 75 meters (250 feet), with most of that possibly occurring within a time scale of centuries. Methane hydrates are likely to be more extensive and vulnerable now than they were in the early Cenozoic. It is difficult to imagine how the methane hydrates could survive, once the ocean has had time to warm. In that event a PETM-like warming could be added on top of the fossil fuel warming.

After the ice is gone, would Earth proceed to the Venus syndrome, a runaway greenhouse effect that would destroy all life on the planet, perhaps permanently? While that is difficult to say based on present information, I've come to conclude that if we burn all reserves of oil, gas, and coal, there is a substantial chance we will initiate the runaway greenhouse. If we also burn the tar sands and tar shale, I believe the Venus syndrome is a dead certainty.

CHAPTER 11

Storms of My Grandchildren

JAKE IS OUR NEWEST GRANDCHILD, our son Erik's first child. Jake has not done much of anything to contribute to global warming. When I snapped this photo of him (figure 31) last year he wasn't even walking yet. He was crawling across the floor and looked up at me when I called his name.

Jake, two years old now, is full of remarkable bubbling optimism and energy. Anniek and I spent a week this summer at the shore with our children and grandchildren. After Jake went to bed at seven P.M. each evening, we could hear him over the monitor his parents use,

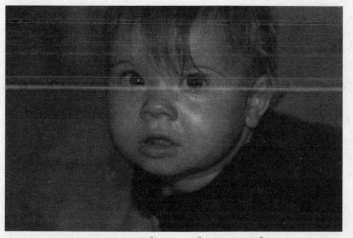

FIGURE 31. *Jake, age eleven months.*

babbling happily for an hour or so before going to sleep—mostly single words and names—Sophie, Conya (for Connor), Oma, Bopa. His mother, Yvonne, a psychologist, says that he enjoys reliving his day before falling asleep. Then he sleeps eleven hours, and it begins all over the next morning.

My parents lived to be almost ninety years old, so Jake may be around for the rest of this century. Jake has no idea what he is in for—that's just as well. He had better first grow up strong and smart.

Over the past few years I thought about our grandchildren and the intergenerational inequity of human-made climate change. Larry King's comment that "nobody cares about fifty years from now" didn't seem right—people do care about their children and grand-children. In fact, the concept of responsibility to future generations is as familiar to Americans as their Constitution, with its phrase "to ourselves and our Posterity" embedded in the preamble. I be-lieved then, and believe now, that if the public had a better under-standing of the climate crisis, they would do what needed to be done.

Year by year I began to make greater efforts to make clear the im-plications of climate science for the public, especially young people. In 2007 I started sending occasional communications about climate change to those on my e-mail list. The list started with several hun-dred scientists, but it grew as other people asked to be added—or I just decided to add them, as in the case of the top two utility com-missioners of every state. My communications begin with a simple instruction for how to be removed from the list, and when a utility commissioner asked to be removed, I would add the next commis-sioner from that state. I wanted such people to understand that a strategic solution to the climate problem requires a phaseout of coal emissions—and consideration of young people demands it.

One of my early e-mail messages was titled "Old King Coal." It was stimulated by a visit to my hometown, Denison, Iowa, where I gave a high school commencement talk—my younger brother Lloyd's son Sam was graduating. The next day I drove with my younger sister, Pat, to Galland's Grove, to the grave sites of our parents. All along the railroad tracks beside our eighteen-mile route from Denison to Dunlap on Highway 30 we saw trains parked back-to-back. I don't know why they were stopped, but what struck me was that about half of the railroad cars in this long string were coal cars. The previous year I had started connecting the dots between global warming and species extinction, based on both the history of Earth and the current unusual rate at which climate zones are shifting.

Most coal trains are long, about one hundred carloads each. A large power plant can burn that amount of coal in one day. The Iowa coal trains made me wonder about the role of coal-fired power plants in the extermination of species. If we continue business-as-usual fossil fuel use, a conservative estimate is that by the end of the century we will have committed to extinction at least 20 percent of Earth's species, that is, about two million species. Based on the proportion of twenty-first-century carbon dioxide emissions provided by one large coal-fired power plant over its lifetime, I concluded that a single power plant should be assigned responsibility for exterminating about four hundred species, even though of course we cannot assign specific species to a specific power plant. Later, in 2008, I cited this conclusion when I testified in defense of activists who had shut down a large coal-fired power plant, Kingsnorth, in the United Kingdom. But it was that day in Iowa when we visited our parents' grave sites that I realized those coal trains are death trains. The railroad cars may as well be loaded with the species themselves, carrying them to their extermination.

But the climate and species story does not need to be one of gloom and doom. As E. O. Wilson, the Harvard biologist, explains, there is a potential path in which the species we have today would survive. Wilson suggests that the twenty-first century will be a "bottleneck" for species, because of extreme stresses, especially climate change. However, if we stabilize climate by moving to energy sources beyond fossil fuels, and if the human population begins to decline as developing nations follow the path of developed nations to lower fertility rates, then a brighter future is possible, a future in which we learn to live with other species in a sustainable way.

This brighter future depends on recognizing what is needed to stabilize climate. My "Old King Coal" e-mail, sent on July 6, 2007, explained the imperative for a moratorium on the construction of all coal-fired power plants unless they are equipped with *actual* carbon capture and storage (CCS) systems. I argued that such a moratorium should be the "rallying issue for young people . . . who should be doing whatever is necessary to block construction of dirty (no CCS) coal-fired power plants." I concluded that "our [scientists'] poor communications" with young people greatly contributed to the problem.

The next day I had an opportunity to try to improve communications, as July 7, 2007, was the first Live Earth event, with worldwide concerts organized by Al Gore and his Alliance for Climate Protection. I had agreed to go onstage at the Meadowlands in New Jersey, on the condition that I could take Sophie and Connor. I was to be the interlude between Jon Bon Jovi and the Smashing Pumpkins. It was deafening when we checked in backstage. I asked the stage manager, "Where's Al?" believing that the plan was for us to go onstage with Al Gore for an impromptu discussion. Apparently, I had misunderstood. No, we were to go out alone, after being introduced by Alec Baldwin—and didn't I have a message to put on the teleprompter?

Um, no. As somebody stood at my shoulder, I wrote down something about how phasing out coal is the essential action for stabilizing climate—and how we needed young people to get involved, to wake up older people, to make it happen. I couldn't quite read the teleprompter, because of my cataracts, so I stumbled a bit. I had told Sophie the day before that I would ask her a question about how many of the animals we should try to save. When I asked her onstage, she answered rather softly, "All of them." But then I asked again, and she said loud and clear, "All of them!" She was a big hit. I held the microphone to three-year-old Connor, whom I was holding, and he said, "Me too."

Our daughter Christine also was onstage with us. Afterward, as the four of us were escorted to our seats, Sophie suddenly burst into tears. I had not thought about the stress of such an event on an eight-year-old. One of the young music fans took off his Live Earth headband and gave it to Sophie, and we reminded her how great she had done. The next morning she slept until eleven thirty A.M., the latest she had ever slept. And once school started that fall, she had something to talk about in third grade show-and-tell.

Participants at Live Earth were asked to pledge to support seven things for the sake of the environment and for the reduction of global warming. I would have preferred a focus on two strategic actions: a rapid phaseout of coal use and a gradually rising price on carbon emissions. The problem with asking people to pledge to reduce their fossil fuel use is that even if lots of people do, one effect is reduced demand for the fossil fuel and thus a lower price—making it easier for somebody else to burn. We must have a strategic approach to solve the problem, with governments providing leadership—it is necessary for people to reduce their emissions, but it is not sufficient if the government does not adopt policies that cause much of the fossil fuels to be left in the ground permanently.

One of the Live Earth pledges concerned fighting against new coal-fired power plants "without the capacity to safely trap and store the CO_2." The danger is that this wording can be taken to imply that a "capture-ready" power plant would be okay. "Capture-ready" is an illusion, a fake, designed to get approval for a coal-fired power plant under the pretense that carbon capture will be added later. The fine print in such applications for power plant approval always includes clauses about feasibility, etc. There is not a snowball's chance in Hades of carbon capture and sequestration being added after the fact. The ratepayers, utility commissions, and politicians would never allow the addition of the technology to capture carbon dioxide, transport it, and sequester it, because it would greatly increase utility bills.

I sent a note to Al Gore and his staff asking for clarification of the pledge. Al responded that he meant exactly what I meant—new coal plants should be allowed only if they actually capture and sequester the carbon dioxide. But then I got a message from his assistant saying that the language allowing power plants that could eventually be retrofitted was what had been recommended by their energy experts. They referred me to a report endorsed by a huge number of energy experts. That, I believe, is the problem. The experts, including those at many nonprofit organizations, have been in Washington too long. They are careful to only nudge industry, asking only for what is "politically realistic" rather than what is in the best interests of the public. They will not state clearly what is needed. That is why young people will need to stand up for their rights.

At the end of my e-mail messages I invite criticisms. One criticism of "Old King Coal" was about my statement that young people should be doing "whatever is necessary" to block coal-fired power plants. This, it was suggested, seemed to be incitement to civil disobedience.

That criticism, it seemed to me, had merit. Even though I had come to see that the response all around—at state, national, and international levels—was basically one of greenwash, all conventional avenues for citizens to affect policy had not been exhausted. And I believed the 2008 elections in the United States could be important.

In my next message, "Old King Coal II," I pointed out that action to deal effectively with climate change was practically impossible as long as our lawmakers are heavily under the undue sway of special interests. I recalled the revolutionaries who declared our independence from a prior king—the intrepid early Americans tried hard to devise a constitution and form of government that could guard against the return of despotic governance and subversion of the democratic principle for the sake of the powerful few with special interests.

The question we needed to ask was this: Did the system still work as intended? Or had special interests found a way to obtain undue influence, far out of proportion to one person, one vote? I warned that "the gleam of a new presidency, by itself, is probably fool's gold." It would be necessary, in addition, to vote in a large number of new representatives and senators from many states, replacing incumbents with those committed to urgent, necessary actions, not to greenwash. So I proposed a Declaration of Stewardship that young people, or anyone else, could use to gauge political candidates. By asking a candidate to make a pledge they would also have a mechanism to hold the candidate responsible for actions after the election.

The Declaration of Stewardship, specifically, was a pledge to support (1) a moratorium on coal-fired power plants that do not capture and sequester carbon dioxide; (2) a fair, gradually rising price on carbon emissions; and (3) measures to improve energy efficiency, for example, rewarding utilities and others based on energy and carbon efficiencies, rather than on the amount of energy sold. Although my proposed declaration never really caught on, in 2007 and 2008 young people did become involved in election campaigning in major ways. Young people, I believe, deserve much of the credit for the surge of support that swept Barack Obama to front-runner status and eventually the Democrats to a landslide victory in November 2008. Students that I met at universities were overwhelmingly supportive of Obama, with optimism that "Change" and "Yes, We Can" were more than just slogans. They put in big efforts to get out the vote. Young people clearly have been trying to use the democratic electoral process in the way it was intended.

Although, as we've seen, the historic election of 2008 has had little effect on the business-as-usual ways of Congress, which is haplessly pursuing an ineffectual cap-and-trade system, that story is not yet complete. It is still possible that the executive branch, with leadership from President Obama, could enter the fray and lead the nation and world in a dramatically different direction. That still may be the best hope for young people. Another possibility is that another nation or nations could force the discussions onto a more sensible track.

However, what is clear, independent of the outcome of specific skirmishes in the U.S. Congress and in international negotiations, such as at the December 2009 climate conference in Copenhagen, is that the results will not yield a "solution" to the climate problem. Whether or not agreements are achieved, the forces for business as usual and continued fossil fuel use assure that any such agreements will constitute, at most, minor steps. The real battle by young people for their future is just beginning.

The battle for the planet, for life on the planet, will surely be conducted on multiple fronts. In addition to the electoral process, there is another option for obtaining justice in our democratic system—the judicial branch. Courts can take a longer view than politicians, who may focus on two-, four-, or six-year election cycles. A judge also has time for thoughtful consideration of a complicated issue.

In the summer of 2007 I went back to Iowa to join a ReEnergize Iowa march organized by the Sierra Club. A discussion of the judicial option came up when I met some people at the march who wanted to prevent the construction of a new coal-fired power plant in Marshalltown. I decided to help them by preparing a sixty-page deposition for the Iowa Utilities Board in an accessible question-and-answer format, including charts with explanatory captions. Halting a single power plant would itself be significant for about four hundred species. But I was hoping that if I prepared written testimony that spoke to the issue of intergenerational justice, it could be used more generally in other judicial proceedings where the matter was relevant.

A decision against the Marshalltown power plant should have been a slam dunk for young people and the planet. The utility's justification for building the power plant was to provide electricity for a factory that would make ethanol out of corn, a process that by itself is detrimental to the planet's climate, even without the added coal emissions from the power plant. I returned to Iowa in January 2008 to give oral testimony, but we lost the case by a vote of 2 to 1. The board's rationale seemed to be the possibility that the power plant would create local jobs. (As yet, the plant has not been built, because of the economic downturn and reduced electricity demand.)

While the decision underlined how difficult it will be to solve the carbon dioxide emissions problem on a case-by-case basis, the judicial option deserves more attention. Of course we must recognize that courts are also subject to fossil fuel influence, as illustrated by the fact that Massey Energy spent $3 million to help elect a judge to the West Virginia Supreme Court; the judge then overturned a ruling against that company. But the likelihood that courts are less beholden to the fossil fuel industry than are the other branches of government is a basis for hope.

How will the judicial branch be brought into play? One way is by young people forcing the issue—not an easy route. The following story, based on an e-mail I sent to my distribution list in October 2008, gives a realistic picture of what young people are facing. I have eliminated here the numerous references supporting the statements; the references are available on my Web site.

Obstruction of Justice

"You're Hannah, right?" Hannah Morgan, a twenty-year-old from Appalachia, Virginia, was one of eleven protesters in handcuffs on the morning of September 15, 2008, at the construction site for a coal-fired power plant being built in Wise County, Virginia, by Dominion Power. The handcuffs had been applied by the police, but the questioner, it turned out, was from Dominion Power.

The earlier discussion between the police and the Dominion

man had taken place too far away to be heard by the protesters. But it almost seemed that the police were working for Dominion. Maybe that's the way it works in a company town. Or should I say a company state? Virginia has one of the most greenwashed, coal-blackened governors in the nation.

It seems Hannah had been pegged by Dominion as a "ringleader." She had participated for two years in public meetings and demonstrations against the plan for mountaintop removal, strip-mining, and coal burning, and she had rejected their attempts to either intimidate or bargain.

Bargain? What bargain is possible when Dominion is guaranteed a 14 percent return on its costs, whether the coal plant's power is needed or not. Utility customers have to cough this up, and they aren't given any choice.

The meetings and demonstrations were peaceful. Forty-five thousand signatures against the plant were collected. But money seems to talk louder. Whatever the Dominion man had said to the police, it must have been convincing. Hannah and Kate Rooth were charged with ten more crimes than the other defendants. Their charges included "encouraging or soliciting" others to participate in the action and were topped by "obstruction of justice." Penalty if convicted: up to fourteen years in prison.

"Obstruction of justice"? Is the Orwellian doublespeak in the charge of "obstruction of justice" not apparent?

Executives in the coal and other fossil fuel industries are now aware of the damage that continued coal emissions causes for present and future life on the planet. Yet their response is to promote continued use of coal, and in some cases even encourage contrarians to muddy the issue in the public's mind. Their actions raise issues of ethical responsibility to the young and the unborn, and a question of legal liability, it seems to me.

The governor of neighboring West Virginia asserted that if there were an alternative energy source, there would not be the need to continue mountaintop removal. But coal is not the only potential source of energy in the region. The case has been made that over time wind turbines on the mountaintops could provide more power than coal does, but if the mountaintops are removed for coal mining, the wind quality becomes less useful for power generation. The governor has not taken up the suggestion of using wind instead of coal.

In Wise County the defense case is even stronger than at Kingsnorth in the U.K., because of the demonstrable local effects of strip-mining. Twenty-five percent of Wise County is already devastated by mountaintop removal. Health problems of local residents associated with coal dust have been well documented. Given all this, the peaceful protest of the demonstrators is commendable. They are merely asking business to invest in Appalachia, not destroy it.

I have argued that it is time to "draw a line in the sand" and demand "no new coal plants." I believe we must exert maximum effort to use the democratic process. But what if new electees turn out like the old? We cannot give up. That's why I am now studying Gandhi's concepts of civil resistance.

As for Hannah Morgan et al. and the proposed coal plant, there is no happy ending here, at least not yet. The defense lawyer realized that a trial would be dangerous. An "unfavorable jury pool" made the possibility of prison time real. With fourteen charges against Hannah and Kate, it was unlikely that a jury would find them innocent of all charges. Result: a "B-minus" plea bargain.

Obstruction of justice, indeed. There are other cases against coal protesters that have gone to trial or will go to trial. I mentioned that I testified for the defense in the Kingsnorth trial in Kent, United Kingdom. Six Greenpeace activists had halted the operation of the Kingsnorth coal-fired power station. They were interrupted and arrested before they had completed painting a message intended for Prime Minister Gordon Brown—"Gordon Bin It"—on the smokestack. They were charged with doing sixty-five thousand British pounds' worth of damage and faced possible prison sentences.

It was a trial by local jury, which had nine women, three men. It was a jury that was interested. In the part that I witnessed, the defense lawyer, Mike Schwarz, did a great job. My written deposition, available on my Web site, was long. But Schwarz had marked certain passages, which he would read aloud and then ask me a question, or ask me to read a statement on such and such a page. It may have lasted an hour or more, but you could have heard a pin drop the entire time.

So there was a lot of publicity and euphoria when the jury found the defendants innocent, on the grounds that breaking a law was justified because they were preventing greater damage in the future. However, the euphoria was rather short-lived. The ruling referred only to this specific case—it did not set a precedent. Furthermore, the British government chose to appeal the verdict, rather than use the verdict and public sentiment as a reason to justify rethinking its position.

The United Kingdom, as the nation most responsible on a per capita basis for fossil fuel carbon dioxide in the air today, could set an example by halting construction of any new coal plants and beginning to phase out existing ones. Britain could achieve this via realistic improvements in energy efficiency and increases of renewable energy and nuclear power. But, in fact, Britain is reopening some coal mines. There is not much hope that other nations will take the sort of actions that are needed if the world's heaviest carbon polluter is so obstinate.

Unless there is a sudden change of heart in London, it seems likely that Britain will agree to the cap-and-trade sleight of hand with some specified "goals" for future emissions reductions. It may be counting on the probability that many other nations may also prefer to fake it, sentencing future generations to live with their mess.

A potentially important legal case within the United States, in Utah, may come to trial late in 2009. I refer to the trial of Tim DeChristopher, the University of Utah undergraduate who outbid oil companies at a Bureau of Land Management auction for the right to drill for fossil fuels on public lands. DeChristopher had no funds to pay for those rights, so he has been charged with a crime with a potential for seven years in prison.

DeChristopher's action speaks to the question of whether it makes sense for us, humanity, to go after every last drop of oil and gas in the ground. His action also relates to the nature of the world that DeChristopher and all other young people will live in, and to their future economic well-being. The essence of these matters can be gleaned from figures 22, 23, and 26 (pages 174, 175, and 184).

If we allow energy companies to go after the fossil fuels on public lands, in offshore areas, in the Arctic and Antarctic, then the larger oil and gas reserve estimates in figure 22 become relevant. As shown in figure 23, exploiting these larger reserves would yield an atmospheric carbon dioxide level about 30 parts per million greater than if these marginal fossil fuels were left in the ground. The cost of drawing down atmospheric carbon dioxide by 30 ppm, with an optimistic estimate of $200 per ton of carbon, would be $12 trillion. Even if such technology is developed, climate damage will be incurred during the period before the carbon dioxide would be removed. The willful dumping of this $12 trillion burden on DeChristopher, and on my and your children and grandchildren, provides strong rationale for his action. This is a gross case of intergenerational injustice. We should all strongly support DeChristopher in his case against the U.S. government. The government cannot realistically claim that it is ignorant of the consequences of its action.

Another legal case, in which I am one of the defendants, concerns arrests made at Coal River Mountain, in West Virginia, on June 23, 2009. About thirty of us were arrested, ostensibly for "obstructing, impeding flow of traffic." A guilty verdict conceivably could result in a one-year prison sentence. At the time of my arrest I was reading a statement in front of a Massey Energy facility, the statement being a demand that Massey (1) withdraw its plans to build a coal silo, which would emit tons of coal dust within 300 feet of Marsh Fork Elementary School; (2) fund the building of a new school to replace the one sitting 400 yards downstream of a three-billion-gallon sludge dam; (3) withdraw its permits to blow up Coal River Mountain, which would destroy the mountain's potential for a proposed wind energy project that would provide permanent clean energy and jobs; and (4) halt mountaintop-removal operations, which are destroying the mountains, poisoning water supplies, and increasing the risk of devastating floods.

When I was at Coal River Mountain, I met local resident Larry Gibson, who invited me to drive with him up the mountain to his house. Gibson refuses to sell his property, which includes a two-hundred-year-old cemetery containing scores of his relatives. He has been the target of drive-by shootings—I saw two bullet holes in the side of his house. I hope the FBI is investigating. On the way down the mountain some thick-necked Massey employees gave us a vigorous one-finger salute—but these may have been a minority; others gave a friendly nod as we passed.

Larry mentioned that when Bobby Kennedy Jr. visited his property and looked at the neighboring scalped mountain, he said, "If any foreign nation had done this to us, we would have declared war on them." But this is not being done by a foreign power—rather by a small number of individuals with enormous political sway. And what we have in Washington is coal-fired senators and representatives who serve as their stooges, advocating this abominable mountaintop-removal practice.

Mountaintop-removal mining poisons water supplies and pollutes the air. Giant sludge dams that hold the waste created by washing the coal are an added hazard for local residents. And yet mountaintop removal yields only 7 percent of the coal mined in the United States, less than U.S. coal exports, so the practice could be prohibited without damage to the country's energy supplies. Only a handful of jobs are involved with mountaintop removal—about twenty thousand in all of Appalachia, far fewer than would be provided by clean energy alternatives.

The most useful outcome from our West Virginia trial would be to bring attention to the mountaintop-removal issue and the impacts of climate change on young people. What is most needed is attention from President Obama. Students at nearby Virginia Tech, who first pointed me to the Coal River Mountain case when I gave a talk there in the fall of 2008, must be terribly disappointed in the president. They worked hard to get out the vote for him. They took for granted, given his statements about "a planet in peril," that he would address the blatant case of mountaintop removal and take a fresh, effective approach to the larger matter of climate change. Instead, I am told, he seems to be listening to the political calculations of Rahm Emanuel and David Axelrod. Perhaps he is concerned that the public will not support a principled stand, although all indications are that the public thirsts for that, rather than the usual compromises with special interests. Or perhaps he has been occupied with other matters.

I am optimistic that reason might prevail soon enough to avoid planetary calamity. Yet it is not too difficult to imagine a tragic course, one in which political calculations result in continued kowtowing to the coal industry and the figment of clean coal, and even the development of unconventional fossil fuels. That could bring on the storms of my grandchildren.

Storms of My Grandchildren

Storms. That is the one word that will best characterize twenty-first-century climate, as policy makers continue along their well-trodden path of much talk without a fundamental change of direction. Our grandchildren are in for a rough ride.

The picture, of a dynamic, chaotic climate transition as ice sheets begin to disintegrate, must be painted with little assistance from climate models. Indeed, early climate models suggested a picture that was surely quite wrong.

Primitive global climate models treated the ocean in simple ways and omitted ice sheet dynamics altogether. The consequence in these early models was pervasive large warming at polar latitudes, lesser warming at low latitudes, and a resulting pronounced reduction of the temperature differences (temperature gradients) from equator to pole. The conclusion, therefore, from these climate models was that storms driven by large-scale north-south temperature gradients would be diminished.

Tragically, real-world temperature gradients this century will not be so simple. In the first decade of this century, while the large ice sheets are just beginning to be softened up, we have seen significant increased warming in the high latitudes of the northern hemisphere, especially in central Asia and the Arctic. But once ice sheet disintegration begins in earnest, our grandchildren will live the rest of their lives in a chaotic transition period. The transition period necessarily will last at least several decades, even if methane hydrates kick in and hasten explosive change, because of the large amount of ice involved.

Business-as-usual greenhouse gas emissions, without any doubt, will commit the planet to global warming of a magnitude that will lead eventually to an ice-free planet. An ice-free planet means a sea level rise of about 75 meters (almost 250 feet).

Ice sheet disintegration will not occur overnight. But concepts about the response time of ice sheets that paleoclimate scientists have developed based on Earth's history are misleading. Those ice sheet changes were in response to forcings that changed slowly, over millennia. Ice sheet responses in the past often occurred in fairly rapid pulses, but disintegration of an entire continental-scale ice sheet required more than a thousand years.

Humans are beginning to hammer the climate system with a forcing more than an order of magnitude more powerful than the forcings that nature employed. It will not require millennia for the ice sheets to respond to the human forcing, but the same inertial forces that slowed the natural response will be in play. It requires a lot of energy to melt ice.

Consider ice initially at a temperature of −10 degrees Celsius. To melt one gram of that ice and bring the water up to the average temperature of Earth's surface (about 15 degrees Celsius) requires about 100 calories of energy. Let's put that into units relevant to the planet: Melting enough ice to raise the sea level 1 meter requires an average 9 watt-years of energy over the entire planet. In other words, if the planet is out of energy balance by 1 watt per square meter, it will take the planet nine years to gain enough energy to melt enough ice to raise sea level 1 meter, if all that energy gain goes into melting ice.

Earth at present, averaged over a decade, is out of energy balance, gaining slightly more energy from absorbed sunlight than it radiates to space as heat radiation. The positive energy balance is due to increasing greenhouse gases, mainly carbon dioxide, which is the dominant climate forcing. However, the imbalance is reduced by human-made aerosols, which reflect sunlight to space. And during the past six years, since 2003, the planet's energy imbalance has been small, at least in part because of the diminishing solar irradiance, as the sun has gone into the deepest and longest solar minimum during the period of accurate solar measurements.

Averaged over a decade, Earth's recent energy imbalance is probably about one-half watt per square meter—but we are not measuring ocean temperature well enough to define the imbalance precisely. However, so far only a small fraction of this energy imbalance is being used to melt ice—most of the energy imbalance is warming the ocean. This division of the excess energy between melting the ice and warming the ocean will shift more to ice melt as the ice sheets are softened up by global warming and begin to discharge ice to the ocean more rapidly.

One effect of increased ice discharge will be to cool the neighboring ocean. So far, the cooling effect of ice discharge is relatively small, although extensive ice shelf melt around Antarctica already has a detectable influence on ocean surface temperature. As greenhouse gases increase under business-as-usual emissions, it is inevitable that the ice sheets will begin to discharge ice more rapidly and have a larger cooling effect on nearby ocean regions.

West Antarctica, the most vulnerable ice sheet, will begin to shed ice at a substantial rate as climate change continues. Ice from West Antarctica will probably be the largest contributor to rising sea level in the twenty-first century and keep the ocean surface around Antarctica near the freezing point, similar to present temperatures.

The Greenland ice sheet rests mostly on land above sea level, so it is not as vulnerable to rapid collapse as West Antarctica, but it can lose mass fast enough to influence North Atlantic Ocean surface temperature. Greenland cannot contribute as much to sea level rise as Antarctica, but freshwater from melting Greenland ice can have a huge impact on the North Atlantic region via its effect on the ocean "conveyor" circulation.

Ocean water in the North Atlantic is rather salty, compared with, say, the North Pacific, in part because of the contribution of very salty Mediterranean water that passes through Gibraltar and moves into the North Atlantic. The combination of high salt content and winter cooling causes North Atlantic surface water to become dense enough to sink to the ocean bottom. As that deep water moves south, warmer water at intermediate levels moves north to replace it.

This ocean circulation can be interrupted by the addition of substantial fresh meltwater, because the resulting less salty surface ocean water is not heavy enough to sink. Numerous documented instances in the paleoclimate record indicate that glacial meltwater can shut down the ocean conveyor circulation, causing a cooling in the North Atlantic region. This phenomenon was the basis for the highly unscientific movie *The Day After Tomorrow*, with incredible near-instant cooling in the northern hemisphere. In reality, if there were a shutdown of deepwater formation in response to global warming and ice melt, the cooling would be only a few degrees and limited mainly to the North Atlantic Ocean, with a small downwind effect in Europe that partially balances greenhouse warming there.

In any case, once Greenland starts shedding ice at a substantial rate, the ice will keep the temperature of parts of the North Atlantic relatively cool. If deepwater formation slows down, regional North Atlantic cooling will be enhanced.

Meanwhile, throughout low latitudes, the atmosphere and the ocean surface will be getting warmer and warmer during the twenty-first century. The effects of increased global warming will exacerbate trends that are already apparent, including melting of mountain glaciers, expansion of dry subtropical regions, more intense forest fires, and competition for diminishing freshwater supplies. A warmer atmosphere causes greater desiccation, but at other times and places it can deliver heavier rain and cause larger floods.

Increased warming's greatest impact on storms will occur through its influence on atmospheric water vapor. The amount of water vapor that the air can hold is a strong function of temperature. The fact that atmospheric water vapor increases rapidly with only a small temperature rise is the basis for the runaway greenhouse effect. But the storms of our grandchildren will begin long before the planet approaches the runaway greenhouse effect.

Even without the chaos that disintegrating ice sheets will bring, the strongest storms will become more powerful this century. That statement is true for storm types that are driven by latent heat. That's a big deal, because storms fueled by latent heat include thunderstorms, tornadoes, and tropical storms such as hurricanes and typhoons.

Latent heat is the energy that water vapor acquires when it evaporates from the liquid state or sublimates from ice. To evaporate water requires a lot of energy—more than 500 calories per gram of water at normal atmospheric pressure—which is needed to break the strong forces of attraction between water molecules. When the water vapor condenses, that latent energy is released as heat that is potentially available to fuel a storm.

Not each individual storm fueled by latent heat will be stronger as the world becomes warmer. Just because there is more fuel around does not assure it will be used; instead, each storm's strength depends on specific meteorological circumstances. However, the strongest storms of the future will have greater wind velocities. That's important, because damage caused by winds is a sharp function of wind speed. Just a 10 percent rise in wind speed increases the destructive potential of the wind by about one third.

Because a warmer atmosphere holds more water vapor and thus

has greater latent heat, the strength of the strongest storms will increase as global warming increases. The greater moisture content of the air also increases the amount of rainfall and the magnitude of floods. Already, as we've seen, many places around the world have experienced an unnatural increase of "hundred-year" floods, which are occurring more often than their name would imply. In some places the effect of increased rainfall amounts is exacerbated by deforestation or other human activities that reduce the ability of the surface to retain water.

The strongest hurricanes and other tropical storms will become stronger, because of the increased "fuel" for the storms. The impact of warming on the frequency of tropical storms is more difficult to predict, because hurricane formation depends on various meteorological factors that can change as climate changes. However, one of the requirements for hurricanes is a sufficiently warm sea surface. Thus the region in which tropical storms can form almost surely will expand as sea surface temperatures rise. Some confirmation of that expectation was provided by Cyclone Catarina, which developed wind speeds of 100 miles per hour in the South Atlantic Ocean in March 2004 before it made landfall in southeastern Brazil. It was the first recorded tropical storm in the South Atlantic.

Even thunderstorms can produce great damage. Thunderstorms usually develop where a warm, moist air front collides with a cool air front. As the warm, moist air rises within the cooler surrounding air, water vapor condenses, releasing its latent heat, which fuels and speeds the updraft. The surrounding compensating downdraft is what causes wind damage on the ground. Unstable air masses along a cold front can produce severe thunderstorms, including large supercell storms with wind speeds of 80 miles per hour or more. Such supercells are the principal spawning ground of tornadoes. In addition to direct wind damage, these supercell storms can produce heavy hail and flash floods.

But an increase in maximum storm strength and an expansion of the regions with severe storms—thunderstorms, tornadoes, and tropical storms—are just the beginning of the storm story. As global warming continues, storm effects will ratchet upward in three major ways.

One of these ratchetings will be the development of more powerful and destructive midlatitude or frontal cyclones. Frontal storms will be more powerful, because they depend upon the temperature difference between the cold and warm air masses as well as upon the amount of moisture in the atmosphere behind the warm front. This intensification of frontal cyclones will be an effect of melting ice sheets, once ice sheets begin to disintegrate rapidly enough to keep regional ocean surface temperature from rising as fast as continental temperatures and temperatures at lower latitudes. The most important point is that there will be places and occasions in which the warm air masses will be loaded with far more water vapor than would be the case in a cooler world.

A taste of this ratcheting's future consequences was provided by the cyclonic blizzard, the Superstorm, that hit North America in mid-March 1993. That storm, referred to in some regions as the "Storm of the Century," was caused by a collision of a cold Arctic air mass and a moisture-laden low-pressure air mass from the Gulf of Mexico. A squall line, a line of severe thunderstorms, formed along the frontal boundary, which moved from the Gulf of Mexico over Cuba and Florida and then up the East Coast of the United States. The storm stretched from Central America to Nova Scotia, Canada. Straight-line winds reached hurricane force in the gulf region, and well over 100 miles per hour in Cuba. The squall line produced "thundersnow" (a snowstorm with thunder and lightning) and a blizzard from Texas to Pennsylvania. Birmingham, Alabama, had seventeen inches of snow and parts of Pennsylvania received two to three feet. Ten million people lost electric power, and three hundred people died from the storm.

Yet the '93 Superstorm will readily be eclipsed by storms in the twenty-first century, as the moisture content of low- and midlatitude air increases and coexists with ice-cooled polar air masses. The intensity of frontal cyclones will increase through the twenty-first century as the rate of ice sheet mass loss increases and warming continues to grow at low- and midlatitudes. This first ratcheting, though, will pale in comparison to the effects of the second ratcheting: when ice sheets' rapid disintegration causes a sea level rise measured in meters.

Remarkably precise measurements of Earth's gravitational field by the Gravity Recovery and Climate Experiment (GRACE) reveal that the Greenland ice sheet has been losing mass for the past few years at a rate of about 100 cubic kilometers per year. West

Antarctica is losing mass at a comparable, although somewhat smaller rate. These precise satellite gravity data go back only to 2002, but other, less accurate data suggest that even as recently as the 1990s these ice sheets were much closer to mass balance, i.e., they were neither gaining nor losing mass at a substantial rate.

Thus it seems that a disintegration of the ice sheets has begun, but so far the effect on sea level is moderate. The rate of sea level rise in the past decade, including the effects of mountain glacier melt and the thermal expansion of warming ocean water, has been 3.4 centimeters per decade, i.e. a rate of 34 centimeters (about 14 inches) per century. This rate of sea level rise will grow as global warming increases. Ice shelves that buttress the West Antarctic ice sheet and some portions of the Greenland ice sheet are melting. As the ice shelves disappear, the rate of discharge of icebergs to the ocean is expected to increase. I have already pointed out another process that may hasten the beginning of rapid ice sheet disintegration: heavier summer rainfall, which may occur over portions of the ice sheets because of the existence of warmer, more moisture-laden air.

Ice sheets eventually begin to disintegrate at rates of several meters of sea level per century, even with the slow pace at which natural climate forcings change. But predicting when ice sheet mass loss will accelerate in the twenty-first century is a notoriously difficult "nonlinear" problem. We could "lock in" disastrous sea level rise very soon, that is, create conditions that guarantee its occurrence, but it is likely to be several decades before a rapid sea level rise begins. On the other hand, we have been surprised by how fast some other climate changes have occurred—such as disappearance of Arctic sea ice, expansion of the area of subtropical climate, and melting of mountain glaciers. If methane hydrates released from the deep oceans and tundra begin to contribute substantially to atmospheric methane, if human-made sulfate aerosols decrease rapidly because we clean up pollution, if solar irradiance bounces back soon from its current low point . . . such factors may accelerate climate change. For the moment, the best estimate I can make of when large sea level change will begin is during the lifetime of my grandchildren—or perhaps your children.

With the combination of a higher sea level, even of only a meter or so, and increased storm strength, the consequences of future storms will be horrendous to contemplate. The problems will not be restricted to those places commonly subjected to tropical storms. Other storms with comparable power will affect populations that are one or two orders of magnitude greater than the number of people displaced by Hurricane Katrina, which struck New Orleans and the American Gulf Coast in 2005.

Consider a storm such as the 1991 Halloween Nor'easter. It began as a low-pressure area over Indiana, which moved to the east-northeast into the Atlantic off Canada. There, the low deepened and moved to the east-southeast, but, encountering a blocking ridge in the northern Atlantic, it curved to the west, where it met northward-moving Hurricane Grace. The hurricane, swept aloft by the cold front and absorbed into the circulation of the deep cyclone, added energy to the cyclonic storm. The minimum pressure fell to 972 mbar (millibar is an old unit for atmospheric surface pressure still employed in weather forecasting—the global average for atmospheric pressure at sea level is about 1011 mbar), with sustained winds of 75 miles per hour, making this extratropical system a Category 1 hurricane. A Canadian buoy at 42N, 62W, about two hundred miles off the coast of Nova Scotia, reported wave heights as great as 31 meters (101 feet). Fortunately, the strongest forces remained offshore, although the northeastern United States was hit with a storm tide of 4 meters (13 feet) with an added storm surge of 1.5 meters (5 feet).

Now consider the situation when sea level is even 1 to 2 meters higher, storms are stronger, and atmospheric moisture content is greater. More powerful Nor'easters and hurricanes will hit the East Coast cities along with higher sea levels—it is not a question of whether, only a question of when. Social and economic devastation could be unprecedented. It is not necessary to put the entire island of Manhattan under water to make the city dysfunctional and, given prospects for continuing sea level rise, unsuitable for redevelopment.

Other parts of the world are as vulnerable, if not more so. Consider the North Sea flood of 1953, which affected the coastlines of the Netherlands and England, and to a lesser extent Belgium, Denmark, and France. The flood was caused by the combination of a high spring tide and a storm tide due to a severe European windstorm. The combined tidal surge in the North Sea exceeded 5 meters above mean sea level. About 1,400 square kilometers were flooded in the Netherlands and 1,000 square kilometers in the U.K. In response, the Dutch have built an ambitious flood defense system. The British, too, built improved flood defense systems, including the Thames Barrier to secure central London against a future storm.

When sea level rise reaches a level of meters—and note that there is no "if" about this sea level rise, only a "when," assuming that politicians are allowed to continue their business-as-usual game—these enhanced barriers will eventually prove futile. Indeed, when the barriers are breached, the area and extent of devastation will be unprecedented. Sea level rise will make a mockery of Dutch plans to build floating houses—unless they plan to live on the open sea. The lowlands of northern Europe will no longer be inhabitable.

What about the effect of sea level rise on developing nations? The consequences for a nation such as Bangladesh, with 100 million people living within several meters of sea level, are too overwhelming, so I leave it to your imagination. No doubt you have seen images of the effects of tropical storms on Bangladesh with today's sea level and today's storms. You can imagine too the consequences for island nations that are near sea level. We can only hope that those nations responsible for the changing atmosphere and climate will provide immigration rights and property for the people displaced by the resulting chaos.

The timing of the third ratcheting effect of global warming, the melting of methane hydrates, is as unpredictable as the others. Warning signs are beginning to appear already, with bubbling of methane from melting tundra and from the seafloor on continental shelves. So far the amounts of methane released in this way have been small. The methane hydrates of greatest concern are those in sediments on the ocean floor, because of their great volume. Although estimates of the current amount of methane hydrates range widely, the long cooling trend of the past 50 million years surely has resulted in an accumulation exceeding that which drove the sudden 5- to 9-degree-Celsius global warming that occurred during the Paleocene-Eocene thermal maximum (PETM) about 54 million years ago.

Global ocean circulation reorganized during the PETM, with deep water formation occurring in the Pacific Ocean rather than the North Atlantic, where it occurs today. The flooding of the ocean floor with warmer Pacific Ocean water may have been a key factor in the melting of methane hydrates during the PETM. Could a change of ocean circulation happen again in the near future? Global models of today's climate sometimes have a problem with spurious formation of deep water in the Pacific Ocean, which suggests that it would not take much change in the densities of ocean surface waters to alter the location of deep water formation. The instigation for such a change could be the freshwater additions to both the North Atlantic and Antarctic oceans, after the rate of ice sheet disintegration in both hemispheres has reached high levels. This freshwater, because it is less dense than salty ocean water, would tend to shut off the usual sinking of surface water in both the North Atlantic and circum-Antarctic Oceans, that is, it could stop the formation of both North Atlantic Deep Water and Antarctic Bottom Water.

When deep water formation begins in the Pacific Ocean, the inertia of the climate system, specifically ocean circulation, will be far too great for humans to stop, even if social systems are still in order. Once large sea level rise begins to devastate coastal cities around the world, creating hundreds of millions of refugees, there may be a breakdown of global governance. But regardless of that, if ocean circulation changes, such that warmer Pacific Ocean water begins sinking to the ocean floor and melting methane hydrates, there will be no plausible way for humans to reverse that change of ocean circulation.

While we can't predict the details of short-term human history, changes will be momentous. China, despite its growing economic power, will have great difficulties as hundreds of millions of Chinese are displaced by rising seas. With the submersion of Florida and coastal cities, the United States may be equally stressed. Other nations will face greater or lesser impacts. Given global interdependencies, there may be a threat of collapse of economic and social systems.

Physical science is easier to foresee. While the timing of the three ratcheting effects is difficult to predict, their effects are not. With methane hydrate emissions added on top of those from conventional and unconventional fossil fuels, the future is clear. Diminishing feedbacks that help to keep the magnitude of natural long-term climate changes within bounds, such as the ability of the long-term carbon cycle to limit atmospheric carbon dioxide, will have no time to counter amplifying feedbacks. The huge planetary energy imbalance caused by the high levels of atmospheric carbon dioxide and methane will take care of any remaining ice in a hurry. The planet will quickly get on the Venus Express.

In the Year 2525

When the global warming topic emerged publicly in the 1980s, I assumed that policies would begin to move in a direction to protect the public and future generations. Unfolding reality paints a different picture. Politicians pretend understanding, while ignoring discomfiting implications of the science.

Is it really conceivable that the world will allow squeezing of oil from tar sands, from oil shale, from coal—and go after every last drop of oil in the ground? The popularity of the slogan "drill, baby, drill" in the last election campaign in the United States made me shudder, as it must have other scientists who recognize the threat of all-out fossil fuel exploitation.

What will the world be like if we do go down this route? The science tells us exactly what we could expect to happen on Earth if we continue our business-as-usual exploitation of fossil fuels. I've referred to it earlier: the Venus Syndrome. But how to portray the horror of that devastation in a way beyond graphs and numbers and phrases we have heard before, like "climate disaster"? Even though science fiction isn't my area of expertise, I use the following scenario as a clarion call. I must try to make clear the ultimate consequences, if we push the climate system beyond tipping points, beyond the point of no return.

"It's not Earth! It's not Earth!"

"What do you mean it's not Earth?"

"The whole planet is covered by haze! It can't be Earth. The guidance system must have gone haywire. Maybe it's Venus, but it doesn't look like Venus."

"Calm down, Spud. It has to be Earth. We checked the coordi-

nates as we were slowing down, as we approached the solar system. Mayflower II was on track to the third planet from the Sun, just as it was programmed."

"This can't be the planet we have been studying for the last ten years. It's nothing like it!"

"Focus the viewer on it and put the image on the screen so we can all see it."

"There. It's not the blue marble. The atmosphere is full of a yellowish dust or haze. You can just barely see through to some surface features."

"We're supposed to be looping in over the south pole, right? That must be Antarctica."

"Yes, it seems to have more or less the right shape. It must be Antarctica. But I don't see any ice. What should we do, Pa?"

"We need measurements. Use the polarizing spectrometer so we know what we're looking at."

Mayflower II left Claron almost five centuries ago. The spaceship had seven crew members: five humanlike creatures and two robots, or droids. Mayflower II was carrying the hope, probably the last hope, for the survival of the claronian civilization.

Claron was the only planet in its solar system with life. Life developed on Claron long before it did on Earth, and it is far more advanced, by about half a billion years. For millions of years claronians had searched the skies for other intelligent life, or any life. They had long since concluded that they must be unique, the only intelligent life within range, or at least the only life that had developed electromagnetic technology that would allow interstellar communications.

They had built extremely sensitive radio receivers, with a receiving area of thousands of square kilometers. Yet century after century they came up empty. They poured more and more resources into the search for life. They had good reason.

The star that Claron circled was a fairly standard main sequence star, somewhat bigger and older than Earth's sun. So it was burning its hydrogen faster, and its radius was expanding more and more, as the star moved closer to reaching its Red Giant phase. Claronians knew that their years were numbered. They still had millions of years perhaps, but for a civilization half a billion years old, it seemed like they were down to their last moments.

Life on Claron works pretty much the same as on Earth. Claronians and animals inhale oxygen, which is used in cellular respiration, and exhale carbon dioxide. Plants use the carbon dioxide in photosynthesis and produce oxygen as a waste product.

The claronians are peaceful and cerebral creatures. Their life span is about 150 years. So their concern was not about their individual lives but rather the fate of their civilization. Perhaps this was because they had so much time on their hands to think. Life had become easy after their technology had reached a point that their droids could do all the work—planting and harvesting the crops, construction, cleaning.

For more than 100 million years the claronians had kept their climate stable by steadily increasing the shielding of their planet from the light of their slowly brightening sun. They had long realized the need to keep both the amount of sunlight and the atmospheric composition in the proper ranges for their life processes—they could not reduce the carbon dioxide to make up for a brightening sun. But with their space technology, shielding the sunlight was not difficult. They put reflecting pellets in orbit about their planet and added more pellets as needed to keep the amount of sunlight within the range that they were adapted to.

The problem was that, as their sun expanded into the Red Giant phase, it would swallow Claron. For their civilization to survive, at least one breeding pair would need to escape to another habitable planet. But there was no other habitable planet in their solar system, only two Jupiter-like giant gas-ball planets. Their hope was to find another solar system with a climate more like that of Claron.

They had studied many planets around other stars. Two planets, less than a light-year away, had spectra suggesting plant life. Claronians worked for millions of years to develop their spacefaring capabilities. Eventually they were able to send missions to the two green planets and also to several lifeless planets. The first missions were carried out with droids, which could survive accelerations to hyperspeed and long journeys without life-support systems. The droids found that life on the two green planets had not advanced beyond algaelike slime, perhaps similar to life on Earth a billion years ago.

Many attempts were made to transplant claronian life to both of the green planets and to the lifeless planets. All missions failed.

The closest they had come to success was establishing colonies of claronians on these planets, within space capsules on the surface. The spacecraft had carried claronian eggs and sperm, as well as seeds for plant life. And while the droids had been able to raise and educate several claronians, they could not get other species to thrive, and the colonies soon died out. They were not able to manufacture a livable environment on another planet.

Their failures were no wonder. How could they mimic a life-support system that had taken billions of years to develop on their planet? Life on Claron was as complex as on Earth, with millions of interdependent species.

Then, near the end of the twentieth century, Earth time, claronian society exploded with the news that radio signals had been detected from a distant source. It was not noise. The signals must have emanated from intelligent life at a great distance.

The signals were mid-twentieth-century radio signals from Earth, located about forty light-years from Claron. Overnight, the study of Earth became the principal activity on Claron. Before long, there were more university students in Earth studies than any other subject. Claronian scientists realized, from technical and educational television programs beamed from Earth, that life there worked in basically the same way that it did on Claron.

English began to be taught as a second language on Claron, with studies beginning in middle school. Television shows broadcast from Earth became popular entertainment. Earth news was reported daily, in English, forty years after the events had occurred on Earth. In 2003, claronians were dismayed to learn of the assassination of John F. Kennedy. While the claronian public was becoming as acquainted with Earth goings on as many earthlings had ever been, their government began devoting enormous resources to planning and developing the Mission to Planet Earth.

The distance to Earth was much greater than that of their prior missions. It would be an exceedingly difficult trip, despite their advanced technologies. Forty light-years would take several centuries, even using powerful acceleration to hyperspeed, which claronians could not withstand. Though they had learned to recycle wastes during space travel, it was implausible to carry claronians on a trip of several centuries that would require multiple generations.

Instead, they would use the technology developed for their failed attempts to transplant life to the green planets and the dead planets. The spacecraft would carry droids and frozen claronian eggs and sperm. The droids would be programmed to carry out the fertilization and serve as surrogate parents to the claronian babies, as they had successfully done on prior missions. But this time, it seemed, there would be no need to transplant or create entire life systems, other species and their ecologies, and create a livable environment—which is why Earth was so attractive.

Earth had an enormous number and variety of species, just like Claron—animals, fish, birds, even bees that pollinated plants, and butterflies. But Earth's animals looked quite different and seemed spectacular to the claronians. Finding such a planet was a dream they had had for millions of years. If only they could succeed in getting their civilization to Earth.

The government presented its plan to the public. The spacecraft, dubbed Mayflower II, *would carry two droids plus claronian eggs and sperm. When they were within twenty-five years of Earth arrival, the droids would fertilize the eggs in an attempt to produce five claronians: two male-female couples and one pilot for the spacecraft. The launch and cruise would be on autopilot, but the pilot would be needed for landing on Earth. There would be a male-female pair from each of the two continents on Claron, East Claron and West Claron.*

The claronians would be raised and educated, in English, during the final quarter century of the flight from canned programs that would teach them about Claron and Earth. When they arrived at Earth, the claronians would be early in their childbearing years, with the aim of saving their civilization. They knew that Earth's sun still had about five billion years before reaching its Red Giant phase, so their civilization would be safe for a time that, even to claronians, seemed to be eternity. The spacefarers would not be able to communicate with Claron during the flight. Once they arrived they would set up a transceiver allowing them to send detectable signals—but even then the great distance meant that eighty years would be required for round-trip exchange of information.

Mission preparation took several decades. On the day of the launch, in the 2030s Earth time, they were receiving Earth news from the 1990s. They learned that earthlings were beginning to change their planet's atmosphere and realized that could spell trouble for life on the planet. But it seemed that earthlings understood what was happening, so surely they would take the steps needed to stabilize their climate.

The Mayflower II *launch went off without a hitch, as did the first few centuries of its journey. As planned, the two droids, named Ma and Pa and programmed with claronian parental qualities, became surrogate parents, raising the claronians to young adulthood. Claronians had individual temperaments—and lots of time for interaction during their twenty-five-year upbringings together.*

The pilot, an offspring of top claronian navigators, was high-strung and individualistic. He was physically strong and had been given a specialized technical education. They called him Spud, because of his preference for a potato-like vegetable. The other four claronians had nicknames too, but their official identities were Female-East, Male-East, Female-West, and Male-West.

As Mayflower II *approached Earth, the claronians were confounded by what they saw. They turned to their surrogate parents for advice. "Pa, we are in the last programmed maneuver. We are going into orbit about Earth, but it doesn't look like what we expected.* Mayflower II *is off autopilot, it's now in Spud's hands. What should we do?"*

"What do the measurements show?"

"The temperature seems to be one hundred degrees Celsius!"

"Where are you looking?"

"That should be the Pacific Ocean, near the equator. But I can't see the surface. It is all cloudy and steamy."

"One hundred degrees—that's the boiling point of water on Earth's surface."

"It can't be cloudy everywhere."

Indeed, they found areas where they could see to the ground, mostly in middle latitudes, including North America, Europe, and Asia. All of these areas were dry deserts with blowing sand. The yellow haze above the clouds, all around the planet, they soon learned, was composed of desert dust.

They had enough fuel to maneuver in the solar system, but there was no possibility of returning to Claron. Nor any reason to try.

"Ma, what should we do?"

"We must go to Mars. Venus is even hotter than Earth."

"But Mars is a dead planet. Like the other dead planets. Claronians cannot survive on lifeless planets."

"Perhaps Mars is different now. Our last information, when we left Claron, is a few centuries old. Maybe things have changed. Maybe humans moved all of their life-forms to Mars."

"How could they do that? Humans are primitive. We see how they work. The irrationality in their politics, the dividing lines they draw on maps, the fighting, the starving people, the abuse of animals—they are still barbaric heathens!"

"It is our only chance."

After further discussion, they decided that Ma was right. Spud put Mayflower II on course to Mars. They went into orbit about Mars and circled many times, making measurements. Mars still seemed to have all the properties that it did before their spacecraft left Claron.

Mars remained cold and lifeless. But they observed five constructions that must have been human-made. Flags identified the constructions as Chinese, American, European, Japanese, and Indian. The Chinese camp was the largest. The droids were programmed to communicate in Chinese. But the claronians spoke only English, because that was the language they were taught and used on Mayflower II. So they decided to land at the American base.

The American installation was a good choice. It was uninhabited, like the others, but, very considerately, the Americans had left detailed documentation of the twenty-first century. The documentation provided a full history, the complete story of how everything had gone so wrong on the perfect planet, the planet of ten million species.

It was to be the only "entertainment" for the claronians for as long as they would live. They knew there was no point in trying to squeeze atmospheric gases out of the stones, to try to create an environment for life. It was hopeless. Life is too complex. They had brought with them eggs of some of their favorite animals, and fish, and birds, and even butterflies—but there would be no point in unfreezing them.

"Pa, Ma, what will you do? You do not need an atmosphere to breathe. You can go on when we are dead."

"We will do what we are programmed to do. We will shut down. Then, in the future, if another claronian expedition arrives, they can turn us on."

"Why would another expedition come to this godforsaken dead planet? It's no better than any other dead planet. There is no life here. They can find a dead planet closer to home. We will send a message as soon as Spud has finished setting up the transceiver. They will get the message in forty years. They will know what a dead planet this is, what a dead solar system it is."

Just then they heard a tremendous rumbling—the Mayflower II was taking off. Spud had gone back to the spaceship alone.

"Spud, what the hell are you doing? Where are you going?"

"I finished my job—the transceiver is working—you can send whatever message you like. What's left for me to do—twiddle my thumbs for a hundred years? I'm going to give those bastards a smack!" he cried over the telecom. "There's plenty of fuel to get to Earth and make a real big pop."

"Spud, they are all dead. There can't be anybody left alive on a planet with boiling oceans and scorched deserts."

"It doesn't matter. What else am I going to do? The damned fools. They had the perfect planet, and they blew it."

Ma didn't understand what was happening. "Why is Spud going to Earth? We learned that it is uninhabitable."

"Ma, Spud has lost it."

"Lost it?"

"It's an American expression. Something snapped. He is not rational. It must be a regressive trait of claronians from millions of years ago."

"Give me the coordinates of where the biggest big-shot coal CEO lived. The one who kept talking about 'clean coal' while bribing judges and Congress and pouring out pollution."

"Why blame that CEO?" Male-West said. "The government in Washington was responsible. They were the ones who were elected to look out for the public interest. They accepted the coal money. Then they retired from Congress and accepted even more coal money. When Congress passed a climate bill, they slipped in rules to keep the coal fires burning. It doesn't matter, though; all the top coal mining officials were located in Washington, close to the government."

"Wait a minute, Spud," said Female-East. "The way I heard it, tar sands became even worse than coal. And by the time they started mining tar sands, everybody knew about global warming. If they had not used tar sands, maybe methane hydrates would not have kicked in big-time. Maybe you should aim for Canada."

"I'm not so sure about that. This is Female-West. It was the United States that egged the Canadians to do it. The pipeline project to bring tar sands oil to the United States was approved by the U.S. government, signed in Washington."

"Washington again—damn!"

"You know, I think you may be looking at this superficially. This is Male-East. Why were they burning dirty coal? Everybody knows it's bad stuff. But they needed energy. I would lay the fault more with the antinukes movement. Nuclear power was the one available alternative to coal. Yes, it made sense to put a hold on nuclear power after the accident in Pennsylvania, even though the antinukes protesters greatly exaggerated the effect of that accident. But after things had been checked out, and after it was realized that a hundred thousand people a year were dying from coal, and that coal was putting all the species on the planet in danger, it was time for reassessment. In fact, most of the public favored the safer next-generation nuclear power, but the antinukes people thought they knew better. Their lives were devoted to stopping it— nothing would change their opinion."

"I don't think you can blame the antinukes movement. They were arguing for what they thought was best. Even if they were in the minority, they can argue for their opinion."

"She's right, I think. It's not so simple. It's hard to point at any one villain. Anyone can express their opinion. But people were elected and sent to Washington to do what's best for everyone, after hearing the opinions. They totally screwed it up, though. The democratic system didn't work. Why?"

"Money, that's why. The power companies wanted business as usual, and they paid for it. Even the nonprofit organizations needed money. They all became part of business as usual, so they didn't want to say what was really needed. The only real priority in Washington was keeping the status quo."

"Washington again—damn."

They knew where Spud was headed. Female-East and Female-West went to the station's observatory and trained the American telescope on the yellowed, dusty planet, setting the coordinates for Washington.

They heard Spud's last words. "You fools. You had to take us with you too. Two civilizations." His eyes narrowed and his muscles tightened as he prepared for impact.

"Oh!" cried Female-East. "There was a puff of yellow dust. He must have made a big crater. Do you want to look?"

"No. Let's go down and tell Ma and Pa. They should send a message to Claron. The news on Claron, forty years from now, will not be pleasant."

THE ABOVE SCENARIO—with a devastated, sweltering Earth purged of life—may read like far-fetched science fiction. Yet its central hypothesis is a tragic certainty—continued unfettered burning of all fossil fuels will cause the climate system to pass tipping points, such that we hand our children and grandchildren a dynamic situation that is out of their control.

Spud's frustration and anger are understandable—he was handed a hopeless situation, so all that he could make was a big bang. We, in contrast, still have the opportunity to preserve the remarkable life of our planet, if we begin to act now. We must rally, especially young people, to put pressure on our governments.

The most essential actions are, first, a significant and continually rising price on carbon emissions, as the underpinning for a transformation to eventual carbon-free global energy systems, with collected revenues returned to the public so they have the resources to change their lifestyles accordingly. This is the most important requirement for moving the world to the clean energy future beyond fossil fuels, but a carbon price alone is inadequate.

Second, the public must demand a strategic approach that leaves most fossil carbon in the ground. Specifically, coal emissions must be phased out rapidly, and the horrendously polluting "unconventional" fossil fuels, such as tar sands and oil shale, must be left in the ground.

We must be jolted into recognizing the remarkable world we

inherited from our elders, and our obligation to preserve the planet for future generations. Belief in this obligation is almost universal. Native Americans speak of obligations to the "seventh generation." It is a paradigm of almost all religions and of humanists that Earth, creation, is an intergenerational commons, the fruits and benefits of which should be accessible to every member of every generation.

Afterword

My grandchildren Sophie and Connor and I need your help. We started a project, but we can't complete it by ourselves. I noticed that there do not seem to be as many monarch butterflies now as there were in the 1970s. At that time, Anniek and I could look out on our small piece of land in eastern Pennsylvania and see several monarchs at the same time. Now we are lucky to see one at a time. Although climate change is probably not the biggest threat to monarchs at present, if it were to affect the trees on the mountain in Mexico where they go to roost in the winter, it could soon become a problem for them.

The monarch population may be in decline in part because the milkweed plants that they are dependent on are diminishing as more land is developed and weeds are eliminated. So early this summer Connor, Sophie, and I dug up some milkweeds along the edge of Frogtown Road, where we expected they would get mowed, and we transplanted them along our horse fence. Though it was late in the year for transplanting, about half the plants survived, albeit with some slightly yellowed leaves. We saw one monarch flitting about them and hoped it would lay some eggs.

Monarchs migrate thousands of miles every year, with those east of the Rocky Mountains wintering in Mexico, millions of them gathering on trees in a small region in the mountains. A single butterfly cannot fly the whole distance—it takes at least a couple of generations. A female lays eggs on a milkweed leaf. When an egg hatches, a tiny larva emerges and eats milkweed until it grows out of its skin. It does this several times until it is a one-and-one-half-inch-long caterpillar. Then it finds a twig or horizontal surface from which it hangs by its back legs, curls into a J shape, and sheds its outer skin, miraculously molting into a shiny green-blue chrysalis,

FIGURE 32. *Sophie and Connor, ages nine and four.*

about the size of an acorn. More miraculously, two weeks later the chrysalis darkens, becomes transparent, and splits open, revealing an orange and black butterfly with crinkled wings. After pumping fluid into the wings and letting them dry off, the butterfly flies away to find something to eat and to resume the trip to Mexico, somehow sensing the direction that it must go.

When we returned from our trip to the shore this summer, we were surprised to find a caterpillar on our slightly forlorn-looking milkweeds, then another, and another. A few days afterward, one of them disappeared, but we soon found it hanging upside down from the horse fence in the J shape, and by later that afternoon it had become a shiny green-blue chrysalis hanging by a thread. The next day another caterpillar disappeared, but we found this one too on the horse fence, fifty feet away, already hanging in the J shape.

In September I had an operation to remove my cancerous prostate. Two days later Anniek drove us to our Pennsylvania home, stopping by the barn so I could check on the status of our "budding" monarchs. I shuffled slowly to the horse fence, as the wounds from my surgery were just beginning to heal.

The first chrysalis dangled as a flimsy shell—its former occupant probably on the way to Mexico. The other chrysalis, the one fifty feet down the horse fence, was beginning to darken, and there were a few sparkles of bright color within the shell.

Segments of bright orange and black wings could be discerned the next morning through the translucent chrysalis shell. When Anniek and I returned in the afternoon, we found a brilliant full-sized monarch hanging upside down from the fence, attached by slender black legs, waiting for its dripping-wet wings to dry. I banged two pot lids together to scare off the horses. Their magnificent, gargantuan heads, poking around for their usual carrot treat, could instantly have crushed the life from the remarkable insect.

I took photos of the butterfly to show Sophie and Connor. We had seen only a few monarchs all summer. This one would need to go south quickly—monarchs fly ten to fifteen miles per hour—to stay ahead of the encroaching cold weather and find a mate. Next summer their granddaughter may stop by to lay eggs on milkweeds that we will have ready.

And next year we will do better, planting seeds from the milkweed seed pods that we have collected. But there need to be more milkweed plants all along the flyway to Mexico. You and your children or grandchildren can help us with our project of planting or transplanting milkweed plants. It would be great to see an increase of the monarch population, or at least a survival of the species.

Our monarch project speaks to an important issue raised in chapter 11—the idea that life can be transplanted to another planet if it becomes no longer sustainable here. The notion that humans are so godlike that they could reproduce miraculous and fragile phenomena such as the monarch on another planet, or transplant them there, is patent absurdity. Once a species is gone, it is gone. And if we destroy this planet, we destroy ourselves.

As I gingerly sat down to write this afterword, postsurgery, my attention was drawn to a five-foot-tall poster that Anniek had found at a garage sale several years ago. It is a grainy blowup of an early 1950s newspaper photo, recording a famous moment in the baseball rivalry between the Brooklyn Dodgers and New York Yankees.

In it, the baseball is frozen in the air, ten feet above the ground. Scooter (Yankee shortstop Phil Rizzuto), who does not fill out his baggy uniform, is pulling up in horrified dismay as he realizes he cannot reach the ball. Billy Martin is caught in mid-dash, glove outstretched, hat suspended in the air behind him. Jackie Robinson, already nearing first base with his graceful stride, looks back over his shoulder. The fate of the World Series depends on whether the ball can be captured before it crashes to the ground.

That ball, today, is planet Earth. We have reached the moment when we must make the full-effort dash to capture our precious globe before it crashes and our team—the team of all species on our planet—is destroyed. But for our team, unlike a baseball team, there will be no chance of a comeback, no next season to do better. This truly is our last chance.

How, though, can today be a critical moment when we do not yet observe great changes in climate? As we've seen, the effects of climate change have been limited in the near term because of climate system inertia, but inertia is not a true friend. As amplifying feedbacks begin to drive the climate toward tipping points, that inertia makes it harder to reverse direction.

The ocean, ice sheets, and frozen methane on continental shelves—all have inertia, resisting rapid change. Heat is pouring into the ocean, and ice shelves are starting to melt. Continued emissions growth will surely cause destabilization of at least the West Antarctic ice sheet.

How close we are to destabilizing frozen methane is unclear. There are already signs of an accelerated release of methane from high-latitude tundra and from the larger reservoir on continental shelves. So far the amount of methane released has been small. But if we continue to increase greenhouse gas emissions, the eventual destabilization of large amounts of methane is a near certainty. We must remember that the human-made climate forcing is not coming on just a bit faster than natural forcings of the past; on the contrary, it is a rapid powerful blow, an order of magnitude greater than any natural forcings that we are aware of.

Storms of my grandchildren—when will these hit with full force? Already the air holds more water vapor than it did a few decades ago. The strongest of the storms that derive energy from water vapor—including thunderstorms, tornadoes, and tropical storms—are becoming stronger, and the associated winds and floods are becoming more extreme.

But qualitatively different storms will occur when ice sheet disintegration is large enough to damp high-latitude ocean warming, or even to cause regional ocean cooling, while low latitudes continue to warm. Global chaos will ensue when increasingly violent storminess is combined with sea level rise of a meter and more. Although ice sheet inertia may prevent a large sea level rise before the second half of the century, continued growth of greenhouse gases in the near term will make that result practically inevitable, out of our children's and grandchildren's control.

Several uncertainties will affect the speed at which more obvious climate changes emerge. One is uncertainty about whether and how solar irradiance will change during the next few years and next few decades. As of October 2009, the sun remains in the deepest solar minimum in the period of accurate satellite data, which began in the 1970s. It is conceivable that the sun's energy output will remain low for decades, as it apparently did a few centuries ago, which may have been the largest contributor to the Little Ice Age. But as we've seen, contrary to the fervently voiced opinions of solar-climate aficionados, such continued low irradiance would not cause global cooling and would *not* stop the continued progression of global warming. This does not mean, however, that the solar effect is negligible. Indeed, if the sun pulls out of its current minimum soon, resuming a typical solar cycle, there may be an acceleration of global warming in the next six to eight years. But whatever happens with solar irradiance, the world is going to be warmer during the next decade (the 2010s) than it was in the present decade, just as the present decade is warmer than the 1990s.

The other major uncertainties that will influence how rapidly climate change effects become obvious are the amount of human-made aerosols and the planet's energy imbalance. Aerosols are the biggest source of uncertainty in terms of the overall forcing that humans are applying to the climate system. The planet's energy imbalance is our best single measure of the state of the system, helping us define how much of a change in atmospheric composition is needed to restore climate stability. Both require improved data.

But our imperfect knowledge of these quantities does not imply uncertainty about the direction that global climate is headed—the world is getting warmer, and it will continue to do so during the next few decades. On the other hand, better knowledge of these quantities will help us refine the atmospheric composition target

that we must aim for. We already know that we should reduce atmospheric carbon dioxide to, at most, 350 parts per million.

Key quantities we should watch to assess the status of potential
climate tipping points are (1) the mass balance of the West Antarctic and Greenland ice sheets, including ice shelves and the principal
outlet glaciers of the ice sheets, (2) the percentage of fossil fuel carbon
dioxide emissions that remains in the air, and (3) changes of atmospheric methane. I will provide organization, discussion, and updates
of these and other key quantities at www.columbia.edu/~jeh1.

Here are reasons to focus on these quantities: (1) If the ice sheets
become more mobile, discharging more ice to the ocean, it will
bode ill for both future sea level and storms. (2) The percentage of
fossil fuel carbon dioxide remaining in the air has averaged about
56 percent for decades, the other 44 percent being taken up by the
land and ocean. If the ability of the land or ocean to soak up carbon
decreases, that could cause global warming to accelerate, which
could amplify other feedbacks. (3) Methane is important because of
the possibility of an increasing discharge from frozen methane.

You need to be well informed, to understand these matters, because you cannot count on governments, the people paid to protect
the public, to deal properly and promptly with the climate matter.
The problem with governments is not scientific ability—the
Obama administration, for example, appointed some of the best
scientists in the country to top positions in science and energy. Instead, the government's problem is politics, politics as usual.

U.S. government scientists, at least those at the highest levels,
cannot contradict a position taken by the president. And President
Obama's assertion that he would "listen to" scientists did not mean
that he would not listen, perhaps with even sharper ears, to political
advisers.

When you learn of a lightly publicized agreement with Canada for
a pipeline to carry oil squeezed from tar sands to the United States,
when you learn of approval for plants to squeeze oil from coal, when
the president advocates an ineffectual cap-and-trade approach for
controlling carbon emissions, when our government funnels billions
of dollars to support "clean coal" while treating next-generation nuclear power almost as a pariah, you can recognize right away that our
government is not taking a strategic approach to solve the climate
problem.

The picture has become clear. Our planet, with its remarkable array of life, is in imminent danger of crashing. Yet our politicians are not dashing forward. They hesitate; they hang back.

Therefore it is up to you. You will need to be a protector of your children and grandchildren on this matter. I am sorry to say that your job will be difficult—special interests have been able to subvert our democratic system. But we should not give up on the democratic system—quite the contrary. We must fight for the principle of equal justice.

One suggestion I have for now: Support Bill McKibben and his organization 350.org. It has the most effective and responsible leadership in the public struggle for climate justice. McKibben has done a remarkable job of helping young people get organized.

But as in other struggles for justice against powerful forces, it may be necessary to take to the streets to draw attention to injustice. There are places where action has begun to have some effect. The government in the United Kingdom, for example, may be turning against coal plants that do not capture carbon emission—strong activism there is surely playing a role. There have been some locally effective actions in the United States as well. But overall, results are small in comparison to what is needed. The international community seems to be headed down a path toward inadequate agreements at best. Civil resistance may be our best hope.

It is crucial for all of us, especially young people, to get involved. This book, I hope, has provided some assistance in understanding what policies we need to be fighting for—and why this will be the most urgent fight of our lives.

It is our last chance.

—James Hansen
October 12, 2009

Acknowledgments

I am indebted to longtime colleague and friend Makiko Sato for her artistic eye, physicist's demand for accuracy, and indefatigable work ethic in producing the figures and checking for errors.

Nancy Miller suggested that I write this book. She and her colleagues at Bloomsbury, including George Gibson, Sabrina Farber, Peter Miller, and Mike O'Connor; as well as copy editor, Maureen Klier; and Dan Miller, were exceptionally supportive throughout the publication process, and especially deserve credit for improving the clarity of my writing.

Comments and suggestions on subject matter were provided by many friends and colleagues, including Yurika Arakawa, Bill Blakemore, Tom Blees, Barry Brook, Mark Bowen, Darnell Cain, Evelyn DeJesus, Paulina Essunger, Jane Halbedel, Vernon Haltom, Martin Hedberg, Pushker Kharecha, Charles Komanoff, Chuck Kutscher, Andy Lacis, Bill McKibben, Richard Morgan, Randall Morton, Matt Phillips, Makiko Sato, Robert Schmunk, George Stanford, Larry Travis, Jim Wine, and Maiken Winter.

Konrad Steffen provided the photo of a moulin on Greenland.

Data for several figures were obtained from NOAA Web sites.

I gratefully acknowledge the support of Hal Harvey and the Hewlett Foundation, Gerry Lenfest and the Lenfest Foundation, Lee Wasserman and the Rockefeller Family Fund, Steve Toben and the Flora Family Foundation, and Theodore Waddell and the Charles Evans Hughes Memorial Foundation, which has permitted me to organize and carry out workshops on air pollution and energy and climate, as well as to speak out on these matters of importance to the public.

Most of all I am indebted to Anniek for her love and support in tolerating my inordinate obsessions and helping me make time to write this book—and to Erik, Christine, Yvonne, Chris, Sophie, Connor, and Jake for their love and inspiration.

Key Differences with Contrarians

Table employed in 1998 debates with Richard Lindzen and Pat Michaels.

1. **Observed global warming: real or measurement problem?**
   RICHARD LINDZEN: Since about 1850 "... more likely ... $0.1 \pm 0.3°C$" (*MIT Tech Talk*, 34, no. 7, 1989).
   JAMES HANSEN: Global warming is 0.5–0.75°C in past century, at least ~0.3°C in past 25 years.

2. **Climate sensitivity (equilibrium response to $2 \times CO_2$)**
   LINDZEN: <1°C
   HANSEN: $3 \pm 1°C$

3. **Water vapor feedback**
   LINDZEN: Negative, upper tropospheric water vapor decreases with global warming.
   HANSEN: Positive, upper and lower tropospheric water vapor increase with global warming.

4. **CO_2 contribution to the ~33°C natural greenhouse effect**
   LINDZEN: "Even if all other greenhouse gases (such as carbon dioxide and methane) were to disappear, we would still be left with over 98 percent of the current greenhouse effect." *Cato Review*, Spring 1992, 87–98. "If all CO_2 were removed from the atmosphere, water vapor and clouds would still provide

almost all of the present greenhouse effect." *Research and Exploration* 9, 1993, 191–200.

HANSEN AND ANDY LACIS: Removing CO_2, with water vapor kept fixed, would cool Earth 5–10°C; removing CO_2 and trace gases with water vapor allowed to respond would remove most of the natural greenhouse effect.

5. When will global warming and climate change be obvious?

LINDZEN: "I personally feel that the likelihood over the next century of greenhouse warming reaching magnitudes comparable to natural variability seems small." *MIT Tech Talk*, September 27, 1989.

HANSEN: "With the climatological probability of a hot summer represented by two faces (say painted red) of a six-faced die, judging from our model by the 1990s three or four of the six die faces will be red. It seems to us that this is a sufficient 'loading' of the dice that it will be noticeable to the man in the street." *Journal of Geophysical Research* 93, 1988, 9341–9364.

6. Planetary disequilibrium

LINDZEN: No known stated position, but his view on climate sensitivity implies a near zero planetary disequilibrium.

HANSEN: Earth is out of radiative equilibrium with space by at least approximately 0.5 W/m² (absorbing more energy than it emits). The planetary disequilibrium, or planetary energy imbalance, is the most fundamental measure of the state of the greenhouse effect. It could be measured as the sum of heat storage in the ocean plus energy going into the melting of ice. Existing technology, including very precise measurements of ocean and ice sheet topography, could provide this information.

Global Climate Forcings and Radiative Feedbacks

Climate forcings and feedbacks—a simplified version of the table presented at the Gore-Mikulski roundtable and Climsat workshop.

Climate Forcings	Measurement Method
1. **SOLAR IRRADIANCE**	Solar Irradiance Monitor
2. **GREENHOUSE GASES**	
CO_2, CH_4, N_2O, CFCs	Operational (ground-based)
O_3, Stratospheric H_2O	SAGE*
3. **AEROSOLS**	
Tropospheric	Polarimeter
Stratospheric	SAGE*
4. **SURFACE REFLECTIVITY**	Polarimeter

Radiative Feedbacks	
1. **CLOUDS**	
Cloud Cover, Height, Optical Depth, Particle Size, Water Phase	Interferometer, Polarimeter
2. **WATER VAPOR**	
Lower Troposphere	Interferometer
Upper Troposphere	Interferometer, SAGE*
3. **SEA ICE AND SNOW**	Operational (satellite)

* SAGE, the Stratospheric Aerosol and Gas Experiment, views the sun from a satellite looking through Earth's atmosphere.

Q&A with Bill McKibben, cofounder and global organizer, 350.org

Bill: Jim, more than a dozen nations have set new high-temperature records this year, and we've seen the all-time marks set for Asia (Pakistan at 129 degrees Fahrenheit) and Southeast Asia. Given that the global temperature has "only" gone up about a degree, can you explain how this kind of heat is possible?

Jim: Sure. What we see happening with new record temperatures, both warm and cold, is in good agreement with what we predicted in the 1980s when I testified to Congress about the expected effect of global warming. I used colored dice then to emphasize that global warming would cause the climate dice to be "loaded." Record local daily high temperatures now occur more than twice as often as record daily cold temperatures. The predominance of new record highs over record lows will continue to increase over the next few decades, so the perceptive person should recognize that the climate is changing.

Yes, global average warming is "only" about a degree, but that is actually a lot. During the last major ice age, when New York, Minneapolis, and Seattle were under an ice sheet a mile thick, global average temperature was about 5 degrees colder than it is now. The last time Earth was 2 degrees warmer so much ice melted that sea level was about twenty-five meters (eighty feet) higher than it is today.

We scientists create a communications problem by speaking about average global warming in degrees Celsius. Global warming in degrees Fahrenheit is almost twice as large (exact factor is 1.8) and warming is about twice as much over land (where people live!)

than over ocean. Also, certain regions and times experience bigger changes than others. (So far the United States has been lucky, with smaller average warming than most land areas. There is no reason to think that luck will continue.)

But remember that weather variability, which can be 10 to 20 degrees from day to day, will always be greater than average warming. And weather variability will become even greater in the future, as I explain in the book, if we don't slow down greenhouse gas emissions. If we let warming continue to the point of rapid ice sheet collapse, all hell will break loose. That's the reason for "Storms" in the book title.

Bill: What was the deal with "climategate"—the East Anglia e-mails and IPCC's "Himalayan error"? Much of the public was left with the impression that global warming may be a hoax!

Jim: There was a real hoax, for sure—perpetrated on the public by people who prefer business-as-usual, people who concocted a disinformation campaign. They want the public to think that the science is suspect. Doubt is all they need. Their tactics included swift-boating and character assassination, using e-mails stolen from scientists' computers. They did an effective job. Now policy makers continue to sit on their hands, leaving fossil fuel subsidies in place, allowing fossil fuel companies to call the tune—and the devil with young people and nature.

Yes, the stolen e-mails exposed bad behavior by scientists, notably a reluctance of some scientists to give deniers the input data for global temperature analysis. That allowed global warming deniers to assert that global climate change was "cooked" data. But that assertion is nonsense. The NASA temperature analysis agrees well with the East Anglia results. And the NASA data are all publicly available, as is the computer program that carries out the analysis.

Look at it this way: If anybody could show that the global warming curve was wrong, they would become famous, maybe win a Nobel Prize. All the measurement data are available. So why don't the deniers produce a different result? They know that they cannot, so they resort to theft of e-mails, snipping private comments out of context, and character assassination.

IPCC's "Himalayan error" was another hoax perpetrated on the public. The perpetrators—global warming deniers—did a brilliant job of playing the scientifically obtuse media like a fiddle. Here is how they did it.

IPCC (Intergovernmental Panel on Climate Change) produced a series of thick reports, several thousand pages long. Of course it is possible to identify minor flaws in it—it is inconceivable that some flaws would not exist within those thousands of pages. The task of the deniers was to find a minor flaw or flaws, and then work the media so as to make the public suspicious of the entire report. They did their dirty work masterfully, for weeks continually releasing tidbits about possible flaws or uncertainties in the report, dutifully reported by the media even though none of the tidbits altered conclusions about the significance of global warming.

The biggest flaw that global warming deniers could find in the IPCC reports was a statement that all Himalayan glaciers may disappear by 2035 if greenhouse gas emissions continued to increase. Actually, because of the great altitude and size of Himalayan glaciers, some of them almost surely will survive longer than twenty-five years. The estimate of 2035 for glacier demise was not even in the main IPCC report on the physical climate system, but rather in a less-scrutinized report discussing practical implications of global warming.

Here is the real-world situation: Glaciers are melting rapidly all around the world—in the Rockies, the Andes, the Alps, and the Himalayas. All glaciers in Glacier National Park in the United States will be gone in about twenty-five years if greenhouse gas emissions continue to increase. We will need to rename it Glacier*less* National Park.

Observed rapid loss of glaciers *confirms* global warming—it is not a reason to question it! Glacier loss also shows the importance of global warming. During the dry season about half the water in rivers such as the Indus and Brahmaputra is provided by glacier melt. If the glaciers disappear there will be more spring snowmelt and greater floods, but a dangerous reduction of fresh water in dry seasons. Hundreds of millions of people depend on these rivers for fresh water.

Yet climate change deniers scored a coup by trumpeting that IPCC had made an error, turning scientific evidence on its head.

Melting glaciers, properly a cause for concern, became a propaganda tool to befuddle the public. A capable media would have exposed the trick. Instead the media facilitated it, spreading "news" that the IPCC report was flawed.

IPCC scientists had done a good job of producing a comprehensive report. It is a rather thankless task, on top of their normal jobs, often requiring them to work sixty, eighty, or more hours per week, with no pay for overtime or for working on the IPCC report. Yet they were portrayed as incompetent or, worse, dishonest. Scientists do indeed have deficiencies—especially in communicating with the public and defending themselves against vicious attacks by professional swift-boaters.

The public, at some point, will realize they were hoodwinked by the deniers. The danger is that deniers may succeed in delaying actions to deal with energy and climate. Delay will enrich fossil fuel executives, but it is a great threat to young people and the planet.

Bill: You must be referring to the urgency created by climate tipping points. Is there new information about tipping points?

Jim: Yes. Let me first clarify the essence of the tipping point matter. The reason that climate change is a threat is that it can sneak up on us. By the time people recognize that big changes are under way and begin to take action, the system may already have enough momentum that it will be very difficult, if not impossible, to prevent catastrophic effects—such as disintegration of ice sheets or ecosystem collapse.

The root of the matter is the great inertia of the planet. The ocean is four kilometers (two and a half miles) deep on average and the ice sheets covering Antarctica and Greenland are two to three kilometers (one to two miles) thick. So when we begin to force the system by increasing atmospheric carbon dioxide, these massive systems begin to change only slowly. But that inertia, that slowness of response, is not our friend. Once the ponderous leviathans begin to react, they can be hard to stop.

The ocean and ice sheets work together to create the possibility of a cataclysm—rapid ice sheet collapse, sea level rise, and the powerful storms that I describe in the book. The ocean is the conduit for the energy that drives the process. As we increase atmospheric carbon dioxide, we cause Earth to be out of energy balance—the planet absorbs more energy from the sun than it radiates to space as heat energy. Almost all of that excess energy is going into the ocean, which is slowly getting warmer.

A warmer ocean affects the ice sheets in two ways. First, and most important, it causes ice shelves to melt. Ice shelves are the tongues of ice protruding from the ice sheets into the ocean. The ice shelves buttress the giant ice sheets, so as the ice shelves disappear the ice sheets can move more freely toward the ocean. Because of their great thickness, ice sheets are "plastic"; they move under their own weight and discharge giant icebergs to the ocean. Second, the warming ocean melts sea ice around the ice sheets, warms the air, and increases the amount of water vapor in the air. These changes increase snow melt on ice sheets and summer rainfall. Running water on the ice sheets causes them to break up faster.

Bill: So what is the new information?

Jim: It is observations of both the ocean and the ice sheets. Data records are becoming longer and more accurate, helping us verify ocean heat uptake and providing an accurate picture of how ice sheets are responding.

Observations of how the ocean's heat content is changing are the main data that we need to determine the planet's energy balance. Is there more energy coming in than going out? We also must account for energy being used to melt the ice and warm the air and ground, but those are smaller terms.

When I wrote the book I was confident that the planet was out of balance by at least half a watt per square meter, more energy coming in than going out. Our climate model predicted an imbalance of about three quarters of a watt, averaged over the ten-to-twelve-year sunspot cycle, with an uncertainty of about a quarter of a watt. But observations at that time were inconclusive—some researchers claimed that the ocean had not been gaining heat in recent years.

Ocean heat measurements have been improving in the past few years. There are now more than two thousand floats distributed around the world ocean. The floats have instrument packages that yo-yo down to depths as great as two kilometers while measuring the ocean's temperature, among other things. Analysis of these data by Karina von Schuckmann and colleagues at the French government space agency reveals that the internal ocean temperature is rising at a rate that shows Earth has been out of energy balance by at least a half a watt per square meter during the past five years, 2005 to 2009, the time of minimum solar irradiance. I expect data over a full solar cycle and the full depth of the ocean to reveal an imbalance closer to three quarters of a watt. But the data already have important implications.

This energy imbalance is the smoking gun for the human-caused greenhouse effect. If climate were just fluctuating, instead of being driven by a forcing such as increasing greenhouse gases, there would be no reason for the planet to be out of energy balance. The imbalance also tells us that there is more global warming "in the pipeline," without any further changes of atmospheric composition. And it tells us by how much we must reduce greenhouse gases to restore the planet's balance and stabilize climate.

Bill: Those are surely important conclusions. But what is the relation to tipping points?

Jim: The warming ocean is the main reason that ice shelves are melting around both Antarctica and Greenland. Loss of ice shelves allows the ice sheets to disgorge ice to the ocean more rapidly.

New data on ice sheets is provided by longer records from the gravity satellite, which is measuring Earth's gravitational field with such a high accuracy that we can measure changes in the mass of the Antarctic and Greenland ice sheets.

Greenland is now losing mass at a rate equivalent to about 250 cubic kilometers of ice per year, and Antarctica is losing about half as much mass each year. The rate of ice sheet mass loss has doubled during the present decade. Sea level is now going up at a rate of about 3 centimeters (about 1⅕ inches) per decade. But if ice sheet disintegration continues to double every decade, we will be faced with sea level rise of several meters this century.

IPCC has estimated only modest rates of sea level rise this century, much less than one meter. But IPCC treats sea level change basically as a linear process. It is more realistic, I believe, that ice sheet disintegration will be nonlinear, which is typical of a system that can collapse. If that is correct, and paleoclimate data provides abundant evidence that ice sheets have collapsed in the past, then the time required for the rate of mass loss to double is an important characteristic of the system.

So the acceleration of ice sheet mass loss over the past decade is a reason for concern, but the record is not long enough. Precise gravity measurements must be continued so we can see whether the increased mass loss was only a blip in the record or whether mass loss continues to accelerate. If the mass loss from the ice sheets doubles again in the coming decade, it will suggest that we are passing a tipping point—and that ice loss and sea level rise may continue to increase more and more rapidly.

Bill: Can we stop that process? Do we understand what is needed to stabilize the situation?

Jim: We can estimate what is needed pretty well. Stabilizing climate requires, to first order, that we restore Earth's energy balance. If the planet once again radiates as much energy to space as it absorbs from the sun, there no longer will be a drive causing the planet to get warmer. Restoring planetary energy balance would not immediately stop sea level rise, but it should keep sea level rise small. Restoring energy balance also would prevent climate change from becoming a huge force for species extinction and ecosystem collapse.

We can accurately calculate how Earth's energy balance will change if we reduce long-lived greenhouse gases such as carbon dioxide. We would need to reduce carbon dioxide by 35 to 40 ppm (parts per million) to increase Earth's heat radiation to space by one-half watt, if other long-lived gases stay the same as today. That reduction would make atmospheric carbon dioxide amount to about 350 ppm.

Bill: Is that how you came up with the policy goal of 350 ppm?

Jim: It is one of several reasons, as we explained in our 2008 "Target CO_2" paper. For example, there is also ocean acidification. As atmospheric carbon dioxide increases, the ocean becomes relatively more acidic. Ocean biologists conclude that for the sake of life in the ocean we need to aim for an atmospheric carbon dioxide amount no higher than 350 ppm.

But yes, Earth's energy balance is indeed the criterion that provides the most fundamental constraint for what must be done to stabilize climate.

Bill: The 350.org team has met opposition from some climate activists who demand an even lower target for CO_2, say 300 ppm or the preindustrial CO_2 amount, 280 ppm. Would the preindustrial CO_2 amount be a reasonable target?

Jim: All that we can say for sure now is that the target should be "less than 350 ppm." And that is all that is needed for policy purposes. That target tells us that we must rapidly phase out coal emissions, leave unconventional fossil fuels in the ground, and not go after the last drops of oil and gas. In other words, we must move as quickly as possible to the post–fossil fuel era of clean energies.

Getting back to 350 ppm will be difficult and will take time. By the time we get back to 350 ppm, we will know a lot more and we will be able to be more specific about what "less than 350 ppm" means. By then we should be measuring Earth's energy balance very accurately. We will know whether the planet is back in energy balance and we will be able to see whether climate is stabilizing.

The reason that we cannot specify now an exact eventual value for CO_2 is because CO_2 is only one of the human-made climate forcings. Humans have also increased the amount of methane and tropospheric ozone in the air—but these gases are short-lived, so if we reduce the sources of these gases the amount in the air will decrease. It is plausible to reduce the amounts of methane and tropospheric ozone and there are good reasons to do so because ozone in the lower atmosphere is harmful to human health and crops. Realistic ozone and methane reductions will alleviate somewhat the amount by which we must reduce CO_2. On the other hand, we expect that humanity will have some success during the next few decades in reducing atmospheric aerosols (fine particles in the air). Atmospheric aerosols are a health hazard, but they have a cooling effect on climate. Reducing atmospheric aerosols will increase the amount by which we must reduce CO_2. However, human-made aerosols will not return to the preindustrial amount in the foreseeable future, nor will the human-made increase of the planet's surface albedo, which also has a cooling effect.

Therefore, it is foolish to demand that policy makers reduce CO_2 to 280 ppm. Indeed, if, with a magic wand, we reduced CO_2 from today's 389 ppm to 280 ppm that change would increase Earth's heat radiation to space by almost 2 watts (per square meter). The planet would rapidly move toward a colder climate, probably colder than the Little Ice Age. Whoever wielded the magic wand might receive a Middle Ages punishment, such as being drawn and quartered.

Bill: Speaking of punishments, you were arrested near Coal River Mountain in West Virginia for protesting against the leveling of mountaintops to extract coal. What was that about, and what is the status?

Jim: Still no trial date has been set. According to the law, I could get as much as one year in prison. I am beginning to think that the authorities do not want a trial.

I was drawn into the mountaintop-removal plight when I gave a talk at Virginia Tech. The students told me about nearby Coal River Mountain, which Massey Energy plans to decapitate to extract coal. Mountaintop removal is morally indefensible. It pollutes the water supply and spoils the environment forever, all for a small amount of coal. Windmills on Coal River Mountain could provide as much energy in about a century. But mountaintop removal will lower the peak about four hundred feet, making Coal River Mountain an ineffectual source of wind energy. Mountaintop removal provides only 7 percent of United States coal production, which is less than the amount that we export. So it cannot be argued that it is needed in order to keep the lights on—it is needed only to line the pockets of a few fat-cat coal executives.

I went to Coal River Mountain to help draw attention to both mountaintop removal and the bigger issue, the need to phase out coal and stabilize climate. I was arrested while standing by the side of the road in front of the Massey Energy offices, reading a statement that Massey should provide funding for a new elementary school, because they had built a huge sludge pond on the side of the mountain right above Marsh Fork elementary school. If that earthen dam breaks, the school could be buried. It seems that Massey is pretty cavalier with the lives of children as well as the lives of miners working for the company.

Despite the publicity, mountaintop removal continues. I am disappointed that the Obama administration has not simply banned mountaintop removal. They could justify that action on environmental grounds. The jobs and economic stimulus from energy alternatives—energy efficiency, renewable energy, and nuclear power—are superior to the kind of jobs and the dirty energy production that is provided by the coal industry. The number of coal jobs has dwindled. Shoving mountaintops into valleys with bulldozers does not require many people.

Bill: Does this indicate that civil disobedience is not useful for solving the climate problem?

Jim: I call it peaceful civil resistance. True, it has failed to achieve the actions needed to solve the climate problem—but every other approach has also failed. Civil resistance is a necessary part of the solution but, by itself, it is too weak as a tool for change.

Bill: Then what else is required?

Jim: The courts, the judiciary branch of government. The courts are less influenced by fossil fuel money than the legislative and executive branches. The situation is analogous to that of civil rights several decades ago. Nonviolent sit-ins drew attention to the immorality of discrimination and helped to get the courts involved. That opened the door to real progress because courts had the ability to order desegregation under the equal protection provision of the Constitution. Eventually lawmakers became involved. Civil resistance was important because it helped broaden public awareness, and high public interest in turn helps to induce judiciary involvement.

What has become crystal clear is that the executive and legislative branches of the government are not going to solve the climate problem on their own. A few years ago I thought that governments may not understand what the science is telling us, the urgency of the matter. But I learned in my interactions with governments in several nations that the governments are not ignorant of the climate problem, they are not unaware of the need to move on promptly to clean energies. Yet at most they set goals and take baby steps because they are under the strong influence of fossil fuel interests. There are too many people profiting from our addiction to fossil fuels— and they have a huge influence on our governments.

Look at what happened in Congress in 2010. The bills that Congress considered were grossly inadequate. The proposed emission reductions were much less than what the science calls for. Also, the bills were full of loopholes and giveaways to the fossil fuel industry, guaranteeing continued reliance on fossil fuels. Nor did the president distinguish himself. The president did not make specific proposals or weigh in with the authority of his office. He should have spoken to the public and demanded that Congress take the actions that are needed for the public interest.

Congress and the president are thumbing their noses at young people. Their failure to act means that young people can look forward to climate deteriorating out of their control, a planet that is much more desolate than the one that we inherited from our parents. My grandchildren, the most recent born just four months ago, probably will be alive for most of this century—my parents lived for almost ninety years. My children and grandchildren will experience the effect of our emissions—they will pay for our profligacy.

The attitude of Congress and the president angers me. They think they can do, or not do, whatever they please. It is as if they have no obligations to young people. Their primary concern seems to be their re-election; how they can beat the other party, make the other party look bad. When the public throws out one party, the other one is little different—they also think they can do whatever they please.

Bill: You have argued that we need a third party, but the nature of our Constitution and the electoral system make it very difficult for a third party to succeed. We don't have time to build a third party movement, do we?

Jim: Probably not. We must force the present government to do its job. Politicians are not free to do whatever they darned well please. They have obligations to young people.

Responsibility to future generations is a concept common to most cultures, as I discuss in the book. Native Americans refer to an obligation to "the seventh generation." Thomas Jefferson wrote that "Earth belongs in usufruct to the living," meaning that we have the right to use property belonging to future generations, but not the right to damage that property. Jefferson, a farmer, used the usufruct concept specifically with regard to the soil, which, he argued, we must not deplete. He did not explicitly discuss the atmosphere, which seemed so huge to the colonials that they never worried that humans might deplete the atmosphere's ability to sustain our lives and livelihoods.

Obligations to young people, it seems to me, are already clear in the second sentence of the Declaration of Independence, "We hold these truths to be self-evident, that all men are created equal, that they are endowed by their Creator with certain unalienable Rights, that among these are Life, Liberty and the pursuit of Happiness." This basic tenet leads directly to the right to equal protection of the laws.

The Fourteenth Amendment of the Constitution declares: "No State shall make or enforce any law which shall abridge the privileges or immunities of citizens of the United States; nor shall any State deprive any person of life, liberty, or property, without due process of law; nor deny to any person within its jurisdiction the equal protection of the laws." Over time the courts ruled that "any person" includes minorities and women, for example, and equal protection provided the principal basis for extension of civil rights to minorities.

Human-made climate change now raises a moral issue as momentous as any that the courts have considered in the past. Today's adults are reaping the benefits of burning fossil fuels while leaving the consequences to be borne by young people and future generations. Are my grandchildren, and other young people, included in the category of "any person" and thus deserving equal protection of the laws? A positive answer, I believe, is obvious.

Bill: You are suggesting that we file suit against the government?

Jim: Precisely so. Begging Congress to be responsible does not work. Exhorting the president to be Churchillian does not work.

On the contrary, Congress has passed laws and the executive branch has defined and carried out policies that trample on the future of young people. Consider the subsidies of fossil fuels and the permission that is given to the fossil fuel industry to use the atmosphere as an open sewer without charge. We cannot let the government pretend that it does not realize the consequences of its actions.

A basis for suing the government is described by legal scholars such as Mary Wood at the University of Oregon. She shows that the Constitution implies a fiduciary responsibility of governments to protect the rights of the young and the unborn. She describes what she calls atmospheric trust litigation. Suits could and should be brought against not only the federal government but also state governments, and perhaps lower levels—and in other nations as well as the United States.

Courts ordered desegregation to achieve civil rights of minorities. Similarly, if a court finds that a government is failing in its obligations to young people, the court can require that government to submit plans for how it will reduce its emissions. Courts have authority to require governments to report back at intervals on the success of their actions and to define corrective actions if they fail to achieve specified reduction.

So we must define the emissions trajectory needed to avoid dangerous human-made climate change. In other words, how fast must emissions decline to avoid passing tipping points with disastrous consequences? I am working with Pushker Kharecha and Makiko Sato to define the required emissions scenario. Our paper will be titled "Sophie, Connor, Jake and Lauren versus Obama and the United States Congress." Although we have not completed that task, it is clear that the requirement will be an annual emissions reduction of several percent per year.

Bill: Wow. Let's say the court instructs the government to reduce emissions so as to yield a safe level of greenhouse gases, which would mean getting carbon dioxide back below 350 ppm. Is it practical to achieve such a scenario?

Jim: Absolutely. But it requires the government to be honest about what is needed. They cannot use tricks such as those in the House and Senate energy/climate bills. Science demands actual reductions in fossil fuel emissions, not phony offsets. An inadequate plan will be quickly exposed by emissions data—the amount of coal, oil, and gas being burned is well documented.

A court would not be expected to mandate *how* emission reductions are to be achieved. The legislative and executive branches are responsible for defining and implementing the laws. But the laws must yield "equal protection." That requirement will force the government to face up to facts. The most fundamental energy fact is this: As long as fossil fuels are the cheapest energy, they will continue to be used.

Fossil fuels are cheapest only because of government policies. First, there are substantial direct and indirect subsidies of fossil fuels. Second, fossil fuel companies are not made to pay for the damage that fossil fuels do to human health. Instead, the public is forced to bear the costs of air and water pollution. Third, fossil fuel companies are not made to pay for the costs of damage to the environment and the well-being of future generations caused by climate change.

The government must face the fact that fossil fuel use will not decline rapidly unless a rising fee is added to fossil fuels, a fee that should be collected from fossil fuel companies at the source before the first sale. Such a carbon fee will be passed on to consumers in the form of higher prices for fossil fuels. Therefore it is important that 100 percent of the collected funds be distributed to the public, preferably as a monthly "green check," although the funds could be used in part to reduce taxes. This "fee and green check" approach would leave about 60 percent of the public receiving more from the green check than they would pay in increased energy prices. The objective is to reward people who reduce their carbon footprint and to stimulate the development of clean energies.

Bill: There are people who say that, in principle, your idea for a fee and green check is the appropriate underlying policy. And if it were accompanied by energy efficiency standards, regulations that remove barriers to efficiency, and appropriate government investments in energy technologies, it would be possible to achieve rapid reduction of carbon emissions. But they say it is unrealistic because in practice Congress always builds in giveaways and favors to special interests, which make the legislation less effective than it should be.

Jim: Sure, that is the way it has worked. But solution of the energy/climate problem requires a different approach. For example, there could be a bipartisan commission that defines appropriate policies to achieve court-ordered emission reductions, with Congress agreeing to either accept or reject the proposed policies without the ability to add in special favors. The public, I believe, is getting really fed up with the government, with the role of special interests and congressional earmarks. If we cannot overcome the role of special interest money in Washington, then both our nation and the planet are in deep doo-doo. This is a crisis, but I believe it is one that we are capable of overcoming.

Bill: There are also a lot of people who say that it doesn't matter what the United States does, because China now has the greatest emissions and its emissions are growing the fastest.

Jim: China is taking the right steps to move toward carbon-free energy. They are now number one in the world in production of clean energy technologies: solar power, wind power, and nuclear power. Also, China stands to suffer greatly from global climate change because China has several hundred million people living near sea level and the country is already experiencing large damaging regional climate disasters.

There is no doubt that China will want to move rapidly toward clean carbon-free energies. When the United States realizes that it must impose an internal fee on carbon emissions, it should not be difficult to get China to agree to do the same.

Also, it is important to recognize that the United States is responsible for three times more of the excess (human-made) carbon dioxide in the air today than any other nation, with China being second. The much greater responsibility for accumulated human-made emissions is true despite the fact that China's population is three times greater than the United States'. So there is no reason to expect China to act first to reduce emissions.

However, there are advantages in beginning to act rapidly. China is investing heavily in clean energies, and it is likely that they will recognize the merits of imposing an internal carbon price to spur development and implementation of clean energies. The United States risks becoming second-class technologically and economically this century if it does not stop subsidizing dirty technologies and instead move toward progressive policies such as fee and green check, which will stimulate development of clean energies.

Selected Sources

More sources, including my communications, presentations, and scholarly publications, are available at http://www.columbia.edu/~jeh1.

Preface

Pool, Robert. "Struggling to Do Science for Society," *Science* 248 (May 11, 1990): 672–73.

Chapter 1: The Vice President's Climate Task Force

Fröhlich, Claus. "Solar Irradiance Variability Since 1978," *Space Science Reviews* 125 (December 8, 2006): 53–65.

Hansen, James, Makiko Sato, and Reto Ruedy. "Radiative Forcing and Climate Response," *Journal of Geophysical Research* 102 (March 27, 1997): 6831–64.

Hansen, James, Makiko Sato, Reto Ruedy, Andrew Lacis, and Valdar Oinas. "Global Warming in the Twenty-First Century: An Alternative Scenario," *Proceedings of the National Academy of Sciences* 97 (August 29, 2000): 9875–80, http://www.pnas.org/content/97/18/9875.full.

Hansen, James, Makiko Sato, Reto Ruedy, Larissa Nazarenko et al. "Efficacy of Climate Forcings," *Journal of Geophysical Research* 110 (September 28, 2005): D18104 (45 pages).

Intergovernmental Panel on Climate Change. *Climate Change 2001: The Scientific Basis*. Edited by John T. Houghton, Yihui Ding, David J. Griggs, Maria Noguer et al. Cambridge: Cambridge University Press, 2001.

———. *Climate Change 2007: The Physical Science Basis*. Edited by Susan Solomon, Dahe Qin, Martin Manning, Melinda Marquis et al. New York: Cambridge University Press, 2007.

Chapter 2: The A-Team and the Secretary's Quandary

Hansen, James. "Case for Vermont" (May 3, 2007), http://www.columbia.edu/~jeh1/2007/Vermont_20070503.pdf.

Hansen, James, Darnell Cain, and Robert Schmunk. "On the Road to Climate Stability: The Parable of the Secretary" (November 14, 2005), http://www.columbia.edu/~jeh1/2005/ateampaper_20051114.pdf.

Chapter 3: A Visit to the White House

Charney, Jule G., Akio Arakawa, D. James Baker, Bert Bolin et al. *Carbon Dioxide and Climate: A Scientific Assessment*. Washington, D.C.: National Academy of Sciences, 1979. http://www.atmos.ucla.edu/~brianpm/charneyreport.html.

Hansen, James. "Can We Defuse the Global Warming Time Bomb?" naturalSCIENCE (August 1, 2003), http://naturalscience.com/ns/articles/01-16/ns jeh.html.

Kerr, Richard A. "Greenhouse Skeptic out in the Cold: A Prominent Meteorologist Says the Greenhouse Warming Will Probably Be a Bust; Experts in and out of the Climate Community Staunchly Disagree with This Latest Iconoclast," *Science* 246 (December 1, 1989): 1118–19.

Petit, J. R., J. Jouzel, D. Raynaud, N. I. Barkov et al. "Climate and Atmospheric History of the Past 420,000 Years from the Vostok Ice Core, Antarctica," *Nature* 399 (June 3, 1999): 429–36.

Siddall, M., E. J. Rohling, A. Almogi-Labin, Ch. Hemleben et al. "Sea-Level Fluctuations During the Last Glacial Cycle," *Nature* 423 (June 19, 2003): 853–58.

Vimeux, Françoise, Kurt M. Cuffey, and Jean Jouzel. "New Insights into Southern Hemisphere Temperature Changes from Vostok Ice Cores Using Deuterium Excess Correction," *Earth and Planetary Science Letters* 203 (2002): 829–43.

Chapter 4: Time Warp

Hansen, James, William Rossow, and Inez Fung, ed. *Long-Term Monitoring of Global Climate Forcings and Feedbacks.* New York: NASA Conference Publication 3234, February 3–4, 1992 (90 pages).

Marshall, Eliot. "Bringing NASA Down to Earth," *Science* 244 (June 16, 1989): 1248–51.

Chapter 5: Dangerous Reticence: A Slippery Slope

Hansen, James. "A Slippery Slope: How Much Global Warming Constitutes 'Dangerous Anthropogenic Interference'?" *Climatic Change* 68 (2005): 269–79.

———. "Defusing the Global Warming Time Bomb," *Scientific American* (March 2004): 68–77.

———. "Scientific Reticence and Sea Level Rise," *Environmental Research Letters* 2 (May 24, 2007): 024002 (6 pages), http://www.iop.org/EJ/article/1748-9326/2/2/024002/erl7_2_024002.html.

Chapter 6: The Faustian Bargain: Humanity's Own Trap

Camp, Charles D., and Ka Kit Tung. "Surface Warming by the Solar Cycle as Revealed by the Composite Mean Difference Projection," *Geophysical Research Letters* 34 (July 18, 2007): L14703 (5 pages).

Hansen, James. "Dangerous Anthropogenic Interference: A Discussion of Humanity's Faustian Climate Bargain and the Payments Coming Due," presentation at the Distinguished Public Lecture Series at the Department of Physics and Astronomy, University of Iowa (October 26, 2004), http://www.columbia.edu/~jeh1/2004/dai_complete_20041026.pdf.

Hansen, James, Reto Ruedy, Jay Glascoe, and Makiko Sato. "GISS Analysis of Surface Temperature Change," *Journal of Geophysical Research* 104 (December 27, 1999): 30,997–31,022.

Hansen, James, Makiko Sato, Reto Ruedy, Andrew Lacis et al. "Forcings and Chaos in Interannual to Decadal Climate Change," *Journal of Geophysical Research* 102 (November 27, 1997): 25, 679–720.

Chapter 7: Is There Still Time?
A Tribute to Charles David Keeling

Bowen, Mark. *Censoring Science: Inside the Political Attack on Dr. James Hansen and the Truth of Global Warming.* New York: Dutton, 2008.

Hansen, James. "Is There Still Time to Avoid 'Dangerous Anthropogenic Interference' with Global Climate? A Tribute to Charles David Keeling," presentation at the American Geophysical Union Fall Meeting, San Francisco, December 6, 2005, http://www.columbia.edu/~jeh1/2005/Keeling_20051206.pdf.

———. "Swift Boating, Stealth Budgeting, and Unitary Executives," Worldwatch Institute (October 15, 2006), http://www.world watch.org/node/4665.

Hansen, James, and Makiko Sato. "Greenhouse Gas Growth Rates," *Proceedings of the National Academy of Sciences* 101 (November 16, 2004): 16109–14, http://www.pnas.org/content/101/46/16109.full.

Hansen, James, Makiko Sato, Reto Ruedy, Pushker Kharecha et al. "Dangerous Human-made Interference with Climate: A GISS ModelE Study," *Atmospheric Chemistry and Physics* 7 (May 7, 2007): 2287–312, http://www.atmos-chem-phys.net/7/2287/2007/acp-7-2287-2007.html.

Chapter 8: Target Carbon Dioxide:
Where Should Humanity Aim?

Hansen, James, Makiko Sato, Pushker Kharecha, David Beerling et al. "Target Atmospheric CO_2: Where Should Humanity Aim?" *Open Atmospheric Science Journal* 2 (2008): 217–31, http://www.bentham.org/open/toascj/openaccess2.htm.

Hearty, Paul, John T. Hollin, A. Conrad Neumann, Michael J. O'Leary, and Malcolm McCulloch. "Global Sea-Level Fluctuations During the Last Interglaciation (MIS 5e)," *Quaternary Science Reviews* 26 (2007): 2090–112.

Zachos, James, Mark Pagani, Lisa Sloan, Ellen Thomas, and Katharina Billups. "Trends, Rhythms, and Aberrations in Global Climate 65 Ma to Present," *Science* 292 (April 27, 2001): 686–93.

Chapter 9: An Honest, Effective Path

Hansen, James. "Dear Prime Minister" (December 19, 2007), http://
www.columbia.edu/~jeh1/mailings/20071219_DearPrimeMinister
.pdf.

———. "Dear Chancellor, Perspective of a Younger Generation"
(January 22, 2008), http://www.columbia.edu/~jeh1/mailings/
2008/20080122_DearChancellor.pdf.

———. "Dear Prime Minister Fukuda" (July 3, 2008), http://www
.columbia.edu/~jeh1/mailings/20080703_DearPrimeMinister
Fukuda.pdf.

Kharecha, Pushker A., Charles F. Kutscher, James E. Hansen, and
Edward Mazria. "Options for Near-Term Phaseout of Coal
Emissions in the United States" draft (June 2009), http://www
.columbia.edu/~jeh1/2009/UScoalphaseout_draft.pdf.

Chapter 10: The Venus Syndrome

Hansen, James. "Climate Threat to the Planet: Implications for
Energy Policy and Intergenerational Justice," Bjerknes Lecture,
American Geophysical Union, San Francisco, December 17, 2008,
http://www.columbia.edu/~jeh1/2008/AGUBjerknes_20081217
.pdf.

Zeebe, Richard E., James C. Zachos, and Gerald R. Dickens. "Car-
bon Dioxide Forcing Alone Insufficient to Explain Paleocene-
Eocene Thermal Maximum Warming," Nature Geoscience 2
(July 13, 2009): 576–80.

Chapter 11: Storms of My Grandchildren

Pritchard, Hamish D., Robert J. Arthern, David G. Vaughan, and
Laura A. Edwards. "Extensive Dynamic Thinning on the Mar-
gins of the Greenland and Antarctic Ice Sheets," Nature 461
(October 15, 2009): 971–75.

Hansen, James. "In Defence of Kingsnorth Six," September 10,
2008, http://www.columbia.edu/~jeh1/mailings/2008/20080910
_Kingsnorth.pdf.

Index

members, 1, 3, 14, 15
second meeting of, 11–15, 32
third and fourth meetings of, 15
"Climate Threat to the Planet"
  (Hansen), 223
Clinton, Bill, 200
Clinton administration
  censoring of science in, 138
  Hansen falling out of favor with,
    130
  and nuclear power, 200
  poor environmental record, 2, 29
clouds
  and climate feedback, 43
  galactic cosmic rays and, 105
coal
  increased use of, 73, 178–79, 183,
    184, 189
  reserves, 173, 174f
coal emissions. *See also* carbon cap-
    ture and sequestration systems
  and atmospheric carbon dioxide, 2
  emissions to date, 174f
  and extinctions, 239
  phasing out
    and atmospheric carbon dioxide
      levels, 174–75, 175f, 184f, 185
    feasibility of, 185–86
    Hansen's efforts toward, 178–81
    impact on climate change, 173,
      176
    necessity of, 172, 179, 180, 239,
      240–41, 269
    strategy for, 204
  pollution effects, 176–77, 196
coal industry
  ethical and legal liability, 245
  exemptions from Clean Air Act,
    217–18
  political influence of, 249, 267
  Waxman-Markey bill and, 215
coal mining, environment damage
    from, 176–77, 184–85, 245–46,
    248–49
Coal River Mountain, West Virginia,
  248–49
coal-to-liquid plants, new construc-
    tion of, 184
Cobb, Robert, 133
Colbert, Stephen, 151
Columbia shuttle disaster, 133

Committee on Oversight and Govern-
    ment Reform, 52
Congress
  antinuclear lobby influence on, 203
  cap-and-trade and, 212
  scientists' testimony before, ad-
    ministration censorship of,
    138–39
  Senate hearings, xv–xvi, 28–29
Connaughton, Jim, 28, 33–34, 52, 53
continental drift
  and carbon dioxide levels, 159–60
  continents, in early Cenozoic era,
    155f, 156
contrarians
  on climate sensitivity, 54, 55–56,
    279
  at Climate Task Force, 10–11,
    11–12
  key differences with (table), xvi, 12,
    14, 279–80
  tactics of, 12, 43, 56, 57–58, 166,
    167–68
Cooney, Philip, 33–34
coral reefs, stresses on, 165–66
Council on Environmental Quality
    (CEQ), 28, 33, 52, 53
  Hansen presentation to (2003), 28,
    33, 34–35, 50–51, 53, 59
  Lindzen presentation to (2003),
    53–54, 56–58
Craig, Larry, 30, 31
crying wolf, negative consequences
    of, 87–88

Daily Kos, 134
"Dangerous Anthropogenic Interfer-
    ence" (Hansen), 98
Davis, Joo, 134
The Day After Tomorrow (film), 252
DeChristopher, Tim, 247–48
Declaration of Stewardship, 242
deforestation, 2, 119, 173
Democratic Party
  exploitation of antinuclear senti-
    ment, 200
  ignoring of fee-and-dividend plan,
    216
democratic process, special interests
    and, 242, 277
denialists. *See* contrarians

Department of Energy
on carbon dioxide caps, 30
commitment to fossil fuels, 75
defunding of Hansen, 88
energy use projections, 21–22, 21*f*
and nuclear power technology,
198, 199
Deutsch, George, 126–27, 128, 129
developing nations
and cap-and-trade plans, 219–20
and fee-and-dividend plans, 221,
222
and sea level rise, 258
Dickens, Gerald, 233–34, 235
diesel engines, emissions standards
for, 51–52
dimethyl sulfide, 43
Domingues, Catia, 101
Dominion Power, 244–46

Earth
as carbon dioxide sink, 117–18,
118*f*, 120, 159–60
as carbon dioxide source, 159–60
early history, freeze-thaw cycling
in, 228–29
as "Goldilocks" planet, 224, 225*f*
life on, impossibility of replacing,
273
orbit and tilt, and glacial-
interglacial climate changes,
47–49
sunlight hitting. *See* solar radiation
thermal radiation from, as function
of wavelength, 62, 62*f*
Earth Observing System, 66–67
earth science community, management of, 68
Earth's energy imbalance
aerosols increases and, 102
carbon dioxide reductions needed
to restore, 165, 166
contrarian *vs.* Hansen position on,
280
and global warming still in
pipeline, 101
and ice sheet melting, 235–36, 251
IPCC estimates of, 102
needed measurements of, 275
net climate forcing and, 100–102,
108–9

oceans' absorption of, 80*f*, 82–83,
101
and pace of global warming, 275
size of, 81–82, 101, 102
"Earth's Energy Imbalance" (Hansen),
113
economic growth, *vs.* emissions reduction, 19–20
economic well-being, and fertility
rates, 221
Eemian period, 37*f*, 39
"Efficacy of Climate Forcings"
(Hansen et al.), 92–93, 105,
226, 227*f*
EIA. *See* Energy Information Administration
Einaudi, Franco, 77, 79
Einstein, Albert, 54, 123
El Chichón, eruption of, 7
election of 2004, 93, 95, 110–11
election of 2008, 242, 243
electric power, sources of
clean coal, 193–94
nuclear power, 194–204
Emanuel, Rahm, 249
emission reduction goals/caps.
See also cap-and-trade plans
ineffectiveness of, 179, 180–81,
182–83, 205
as tax, 212
end-Cretaceous extinction, 150
end-Permian extinction, 147, 148–50
energy. *See also* Earth's energy imbalance; electric power
adequate supply of, *vs.* carbon dioxide emissions reduction, 19–20
use, projections of, 21–22, 21*f*, 207
energy efficiency
in automobiles, A-Team proposals
for, 23–24
cost-effectiveness of, 230–31
in Germany, 188–89
inadequacy of, 191–92, 202
and phasing out of fossil fuels,
190–91
energy experts, political realism of,
241
Energy Information Administration
(EIA), 21–22, 21*f*
energy infrastructure, necessary
changes to, 19, 73, 115

A Note on the Author

Dr. James Hansen is perhaps best known for bringing global warming to the world's attention in the 1980s, when he first testified before Congress. An adjunct professor in the Department of Earth and Environmental Sciences at Columbia University and at Columbia's Earth Institute, and director of the NASA Goddard Institute for Space Studies, he is frequently called to testify before Congress on climate issues. Dr. Hansen's background in both space and earth sciences allows a broad perspective on the status and prospects of our home planet. This is his first book.